M. Tepfer · E. Balázs (Eds.)

Virus-Resistant Transgenic Plants: Potential Ecological Impact

Springer

Berlin
Heidelberg
New York
Barcelona
Budapest
Hongkong
London
Milan
Paris
Santa Clara
Singapore
Tokyo

M. Tepfer E. Balázs (Eds.)

Virus-Resistant Transgenic Plants: Potential Ecological Impact

With 16 Figures and 16 Tables

Springer

INRA EDITIONS

OECD OCDE PARIS

DR. MARK TEPFER
Laboratoire de Biologie Cellulaire
INRA-Versailles
78026 Versailles Cedex
France

DR. ERVIN BALÁZS
Institute for Plant Sciences
Agricultural Biotechnology Center
POB 411
2101 Gödöllö
Hungary

ISBN 3-540-63257-3 Springer-Verlag Berlin Heidelberg New York

CIP-Data applied for
Die Deutsche Bibliothek - CIP-Einheitsaufnahme
Virus resistant transgenic plants : potential ecological impact ; with 16 tables /
Mark Tepfer ; Erwin Balázs (ed.) - Berlin ; Heidelberg ; New York ; Singapore ;
Barcelona ; Budapest ; Hong Kong ; London ; Milan ; Paris ; Santa Clara ; Tokyo :
Springer, 1997
ISBN 3-540-63257-3

Springer-Verlag Berlin Heidelberg New York
a member of BertelsmannSpringer Science+Business Media GmbH

Camera ready by Mark Tepfer, Versailles, France
Cover design: D&P, Heidelberg
Cover Photograph: © E. Balázs, Field test of transgenic PVY-resistant tobacco
SPIN 10787963 31/3111 - 5 4 3 2 - Printed on acid-free paper

Preface

These Proceedings evolved from the OECD Co-operative Research Programme workshop on "Potential ecological impact of transgenic plants expressing viral sequencies", held at the Agricultural Biotechnology Center in Gödöllö, Hungary on 24-26 April 1997.

The OECD Co-operative Research Programme

At the Directorate for Agriculture of the Organisation for Economic Co-operation and Development (OECD) a co-operative research programme for "Biological Resource Management" has existed since 1990. It focuses on work in four specific topic areas, one of which is "Ecology and utilisation of new organisms" (Theme 3). The activities promoted by this programme are post-doctoral fellowships (announced annually), and the organisation of expert workshops (1-2 workshops per Theme per year). The 26 OECD member countries participating in this programme are: Australia, Austria, Belgium, Canada, Czech Republic, Denmark, Finland, France, Germany, Greece, Hungary, Ireland, Italy, Korea, Japan, the Netherlands, New Zealand, Norway, Poland, Portugal, Spain, Sweden, Switzerland, Turkey, the UK, and the USA.

Examples of the items included in Theme 3 of the programme include:
* The introduction of new organisms and transgenic crop plants
* Insect and weed biocontrol with new organisms
* Population biology of new organisms in agricultural systems
* Biodiversity and ecological consequence assessment
* The sustainability of agricultural systems using new organisms
* Mathematical modelling of new organism introductions

Previous recent workshops, organised within Theme 3 and funded by the OECD Programme include:
* Ecology of introduced, exotic wildlife: fundamental and economic aspects (proceedings by Nummi, 1997)
* Introduction of non-endemic nematodes for biological control: scientific and regulatory policy issues (proceedings by Ehlers and Deacon, 1996)
* Transgenic crop plants: agronomic, economic and ecological aspects (proceedings by Hokkanen and Tsaftaris, 1996)

* Benefits and risks of introducing biological control agents (proceedings by Hokkanen and Lynch, 1995)
* Ecological implications of transgenic crop plants containing *Bacillus thuringiensis* toxin genes (proceedings by Hokkanen and Deacon, 1994)

For further information about the OECD Programme please contact Professor Heikki Hokkanen, Head of Programme, Directorate for Food, Agriculture and Fisheries, OECD, 2 rue André-Pascal, F-75775 Paris Cedex 16, France (e-mail: heikki.hokkanen@oecd.org) or consult the OECD Internet site (http://www.oecd.org).

The workshop on transgenic plants expressing viral sequences

The first virus-resistant transgenic plants were described more than a decade ago, when it was shown that transgenic plants expressing the tobacco mosaic virus coat protein were resistant to this virus. Since then, their results have been confirmed in more than a hundred different host/virus combinations, and in several cases plants have already been field tested and even released for commercial use. Along with these efforts, several other alternative protection technologies have been discovered and described. These techniques are based on pathogen genes other than the coat protein gene. Genes encoding the plant viral replicase and the movement protein are also able to induce protection after integrating into the host plant genome. Protection was also reported using the antisense approach or using catalytic RNAs. As viral satellite and defective interfering RNAs can modulate virus-induced disease symptoms, it was quite obvious that they could also be used for plant protection. Indeed, by engineering satellite and DI RNAs, effective protection was obtained against the helper viruses. The above mentioned pathogen-derived resistances have the potential to play a role in integrated pest management.

The rapid development of molecular biological techniques has enabled us to detect minor differences among virus strains, and experimental data suggest that viral recombination is quite frequent in nature. Using coat protein specific antisera, transcapsidation was also shown to be widespread among virus strains. Some of these experiments were carried out using transgenic plants expressing viral sequences, which led to the conclusion by the opponents of genetic engineering that agricultural use of these transgenics could cause unforeseen negative effects on the environment.

Since the domain covered by the OECD Co-operative Research Programme includes studies of the potential ecological impacts of genetically modified organisms, it gave us the impetus to organise this workshop on the potential ecological impact of transgenic plants expressing viral sequences. Our major concern was to convene world-wide those experts who are most actively working in this field, in the expectation that on the basis of their latest results we will be able to separate facts from science fiction, convinced that this will be of use to the decision makers and legislators responsible for the contained use and deliberate release of GMOs.

References

Ehlers R-U, Deacon J (ed) (1996) OECD Workshop on the introduction of non-endemic nematodes for biological control: scientific and regulatory policy issues. Biocontrol Sci Technol, Special Issue, 6 (3) 291-480

Hokkanen HMT, Deacon J (ed) (1994) OECD Workshop on ecological implications of transgenic crop plants containing bacillus thuringiensis toxin genes. Biocontrol Sci Technol, Special Issue, 4 (4):395-607

Hokkanen HMT, Lynch JM (ed) (1995) Biological control: benefits and risks. Cambridge University Press, Cambridge, 304 pp

Hokkanen HMT, Tsaftaris SA (ed) (1996) OECD/MAICH Workshop on multidisciplinary assessment of transgenic crop plants: technical, agronomic and ecological aspects. Field Crops Res, Special Issue, 45(1-3):1-196

Nummi P (ed) (1997) Proceedings from the OECD Workshop on ecology of introduced, exotic wildlife: fundamental and economic aspects. Wildlife Biol, Special Issue, 2(3):129-232

Ervin Balázs
Institute for Plant Sciences
Agricultural Biotechnology Center
POB 411
2101 Gödöllö, Hungary

Mark Tepfer
Laboratoire de Biologie Cellulaire
INRA-Versailles
78026 Versailles Cedex, France

Heikki Hokkanen
Directorate for Food, Agriculture and Fisheries
OECD
2 rue André-Pascal
75775 Paris Cedex 16, France

Contents

Chapter 7 Behavior of Cucumovirus Pseudorecombinant and Recombinant Strains in Solanaceous Hosts

M. Jacquemond, K. Salánki, I. Carrère, E. Balázs and M. Tepfer

Chapter 8 Recombination Between Cauliflower Mosaic Virus and Transgenic Plants that Contain CaMV Transgenes: Influence of Selection Pressure on Isolation of Recombinants

W.M. Wintermantel, L. Király, J. Bourque and J.E. Schoelz

Chapter 9 Synergy of Virus Accumulation and Pathology in Transgenic Plants Expressing Viral Sequences

P. Palukaitis and I.B. Kaplan

Chapter 12 The Potential of a Beneficial Satellite RNA of Cucumber Mosaic Virus to Acquire Deleterious Functions: Nature Versus Greenhouses

D.Gallitelli, F. Grieco and F. Cillo

Chapter 13 Ecological Impact of Transgenic Virus–Resistance in Crop, Weed, and Wild Plant Populations (Due to Potential Alterations of Plant Invasiveness)

D. Bartsch

Chapter 14 Risk Assessment of Gene Flow Associated with the Release of Virus–Resistant Transgenic Crop Plants

M. Fuchs and D. Gonsalves

Virus Evolution; the Past, a Window on the Future?

Mark J. Gibbs[2], John Armstrong[1], Georg F. Weiller[1] and Adrian J Gibbs[1, 2]

[1] Research School of Biological Sciences, Australian National University, GPO Box 475, Canberra ACT 2601, Australia.

[2] Cooperative Research Centre for Plant Science, c/o CSIRO Division of Plant Industry, GPO Box 1600, Canberra ACT 2601, Australia.

Abstract

The study of biological evolution has many goals. One is to reveal the relationships of organisms. Another is to reveal patterns of past evolutionary change, which may be used, with caution, to make predictions; if patterns recur, and there is no evidence that controlling factors have changed, then similar patterns of events may continue in the future. Such predictions will be particularly valuable when planning new practices that could affect the biological *status quo*, such as the genetic transformation of domesticated organisms with phylogenetically disparate, or entirely novel, genes.

Our understanding of evolution has been revolutionized recently by gene sequencing. Viruses have been found to be polyphyletic and to have evolved in a reticulate pattern dominated by genetic recombination. Most major viral lineages (families and genera) have been established by intra-group (heterologous) recombination and the generation of genes *de novo*. The major lineages of plant viruses differ in their biochemical and ecological life styles, whereas isolates of single species usually differ only by point mutations and in the virulence of the symptoms they produce. Homologous recombination may be one of the principal mechanisms generating new viral species with novel host ranges. However, very few of these kinds of significant events have been reproduced experimentally and thus their frequency and the circumstances, and the forces that favour them, are largely unknown. By charting characteristics against phylogenies we can rank the relative likelihoods of various consequences; recombination is more likely to cause a change in symptom severity or host range than in the mode of transmission. Thus, crops that have been genetically transformed with viral genes and that present opportunities for viral recombination with naturally spreading viruses are likely to produce viruses with novel host ranges or ones that produce altered symptoms.

The VIDE database gives information on the relative susceptibility of different plant species and families, and hence which species might be useful for monitoring and detecting novel recombinant virus genotypes.

Prelude

Theodosius Dobzhansky in his book *Heredity and the Nature of Man* (1964) wrote: "Several scientists debated the idea of evolution before Darwin, notably the great

French biologist Lamarck in 1809. It was, however, the genius of Darwin that made the idea of evolution the pivotal generalization of biology. The year 1859, which saw the publication of Darwin's great book *On the Origin of Species,* is one of the turning points not in the history of biology alone, but in the intellectual history of mankind. It gave man a new insight into his own nature, and into that of the world in which he lives. Nothing makes sense in biology except in the light of evolution."

As the "pivotal generalization", the study of evolution has many goals. The principal one is to reveal patterns of past evolutionary change, and hence the relationships of organisms, past and present. One often–ignored reason for studying these patterns is that if recurring patterns of evolutionary change can be found, then they can be used to make sensible predictions of future change, or at least indicate which sorts of change are more likely to occur than others. These predictions will be particularly valuable when planning new practices that could affect the biological *status quo*, such as the genetic transformation of domesticated organisms with phylogenetically disparate, or entirely novel, genes.

Initially evolutionary patterns and relationships were deduced by comparing the phenotypes of organisms, and by studying fossils. Most resulting versions of the 'tree of life' were dominated by the obvious organisms; animals, plants and fungi formed the major branches sprouting from an amorphous trunk of protozoa and bacteria. The use of phenotypic characters also dominated early work on virus evolution and taxonomy. Few characters were available; host specificity and symptoms first and later, the modes of transmission and virion morphology. Using these data, viruses were grouped and the shared characters provided predictions that could be confirmed. However the phylogenetic relationships between strains, species and groups, and the processes that produced the great diversity of virus phenotypes, were not known.

Over the past quarter century, two biological revolutions have completely altered this perspective. Studies of the nucleotide sequences of genes and the amino acid sequences of the proteins they encode, have provided a vastly increased level of understanding of evolutionary processes and knowledge of the relationships of organisms, especially those with phenotypic characters that previously were difficult to interpret. In parallel, genetics, reverse genetics, biochemistry and studies of molecular structure have provided equally significant advances in our understanding of the structure and function of genomes, genes and proteins. Furthermore, two kinds of evolutionary information are now readily recognised in sequences and may be interpreted in terms of structure and function. Conserved patterns usually indicate common ancestry and the maintenance of useful structures and functions by selection and by the restrictions of the genetic code. By contrast, differences between related sequences indicate a level of flexibility in the form and function of a molecule, and testing of this flexibility through mutation and genetic change, permitted by the redundancy of the genetic code. With these new views our understanding of viruses has changed dramatically. Some phenotypic characters are now well understood, such as the virion morphology and the mode of transmission of many viruses. Others, such as the host range, pathogenicity and epidemiology of most viruses remain largely unknown.

Patterns and phylogenies

All cellular organisms possess nuclear ribosomal RNA genes. Comparisons of their sequences have shown that cellular organisms have evolved almost entirely through speciation and divergence to produce a bifurcating tree-like hierarchy of lineages. These analyses have revealed relatively few surprises and have largely confirmed the single 'tree of life' deduced from comparisons of morphology. By contrast, comparable studies of viral genomes, gene sequences and proteins have shown viruses to be polyphyletic, have discovered unexpected groupings of viruses previously thought to be unrelated, and, most importantly, revealed that much of viral evolution is dominated by recombination.

Thus if the history of cellular evolution is best represented as a conventional tree with three main trunks (archaea, eubacteria and eukaryotes), that of the viruses is totally unconventional and resembles a collection of occasionally anastomosing "strangler vines" hanging from that tree.

When the kinds of differences that distinguish different viral taxa are examined, some recurrent patterns emerge. The gene sequence differences are easiest to categorize (Fig. 1). The most closely related isolates differ by point mutations, often silent, but as one compares more and more distantly related isolates one find increasing evidence of non-silent mutational differences and insertions/deletions of increasing size, sometimes involving novel genes. Within most genera of viruses different genes usually have the same phylogenies, indicating that their evolution has been linked, that they have experienced the same speciation events and have co-diverged. However, this is not true when gene and genome comparisons are made between most families, indicating that major inter-group recombinational events have occurred (see below). Recombination has also been shown to occur between closely related virus species.

Surprisingly, viral phenotypic differences show recurrent patterns which can be mapped, albeit in a very general way, onto the phylogenies of genes. Isolates of individual viral species differ most frequently in symptom severity and, only occasionally, infect different host species or are transmitted by different, but related, vector species. Related viral species, by contrast, are usually defined by the fact that they infect different host species, indeed they may have no host species in common, and they are usually transmitted by different species of the same vector family. Finally, viruses of different genera or families usually have different life styles, infect different host tissues, spread between hosts in different ways and produce virions of different morphology. It is these major groupings that seem mostly to have been established by inter-group (presumably heterologous) recombination.

Processes

Evidence of this kind of process is found in the sequences of the *Picornaviridae* and Comoviruses which have homologous RNA dependent-RNA polymerases and proteases but non-homologous virion protein genes (Gibbs, 1987), similarly from

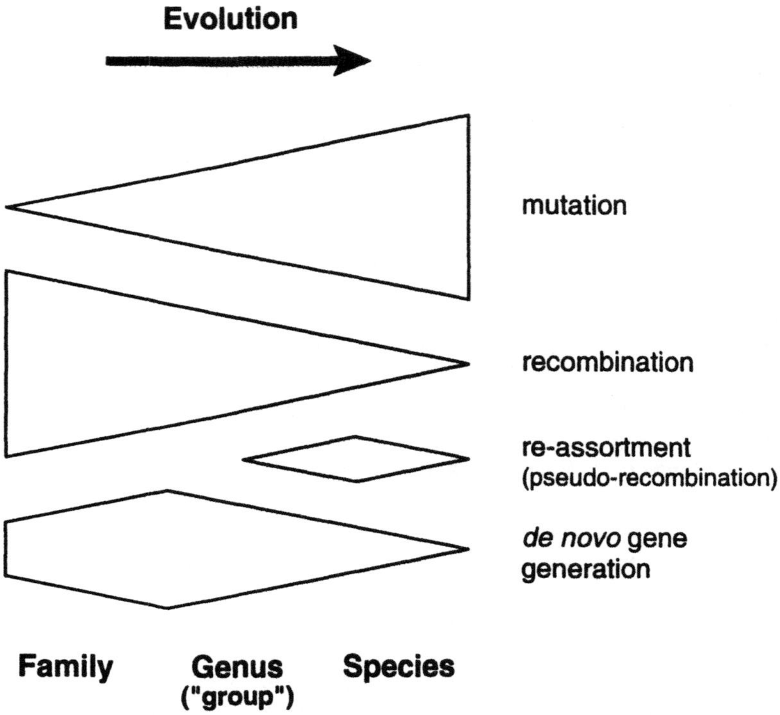

Fig. 1. Diagram indicating the apparent relative importance of various evolutionary processes in the generation of genetic novelty among viruses.

the relationships between luteoviruses (Veidt *et al.*, 1988; Rathjen *et al.*, 1994; Gibbs and Cooper, 1995), between capillo–, carla–, marafi–, potex–, tricho– and tymoviruses (Koonin and Dolja, 1993; Edwards *et al.*, 1997) and between potyviruses and several other genera (Goldbach, 1992). Typically this kind of recombination involves virion protein genes and genes that function in replication. Homology between replication genes from different genera and families suggests they have ancient origins (Gorbalenya, 1992), and the same is true of some virion protein genes. However, there is considerable disagreement on the ages of viral lineages; some believe that viruses are as old as life itself (Kamer and Argos, 1984), and most lineages commensurately ancient, whereas others have used extrapolations from gene sequence comparisons of viruses collected over the past few decades, often from epidemics, to conclude that some viruses have originated in relatively recent times (Zanotto *et al.*, 1996; Weaver, 1995). Hence, although it is impossible to date inter-group recombinational events, relative dates may be estimated, and it is clear that such events occur frequently and are evolutionarily important.

Another class of genes that is common in viral genomes, are those that are shared by different species of the same genus, but which show no sequence similarity to other genes in the international databases. Some of these genes encode ecologically important proteins, such as those that assist cell to cell movement (Deom *et al.*, 1994) or parts of proteins involved in transmission by vectors (Hancock *et al.*, 1995). Although there are some examples (e.g. Gibbs *et al.*, 1996), there appears to be less evidence of recombinational exchange of these kinds of genes between viruses. One likely reason for the lack of evidence is that these genes are evolving more quickly than others in the genome, and the clues of their relationships may have been partially or completely erased. It is also clear that many of these genes arose *de novo* in the genome of a common ancestor of a group (Keese and Gibbs, 1992; Gibbs and Keese, 1995), or have arisen more recently by replication slippage (Hancock *et al.*, 1995). A third possible explanation for the lack of homologues is that some of these genes may have been captured from host genomes by recombination, and their host encoded homologues have yet to be sequenced. There is considerable evidence of host gene capture by viruses that infect vertebrates (e.g. Johnson *et al.*, 1993), but much less sequence evidence of this among plant viruses (Gorbalenya, 1992). This may reflect the fact that many more vertebrate genes have been sequenced than plant genes, that most viruses with DNA genomes probably have smaller mutation rates than those with RNA genomes, and that many viruses of vertebrates have large DNA genomes, whereas most plant viruses have small RNA genomes.

The potential for RNA genomes to evolve rapidly is an important factor in another kind of recombinational process. Viral RNA genomes are able to evolve rapidly because the replicases that copy these genomes frequently mis-read templates and lack proof-reading and repair activities and thus incorporate point mutations into progeny genomes (Domingo *et al.*, 1978; Steinhauer *et al.*, 1992). RNA virus mutation rates are often estimated to be in the order of million fold greater than the mutation rates of their cellular hosts (Domingo and Holland, 1994). For most viruses, selection constrains this propensity for mutational change, but when their ecology changes, especially the acquisition of a new host, and differences arise between populations of a species, intra-species (homologous) recombination permits the exchange of these variant sequences. This kind of exchange is significant in the evolution of lentiviruses (Robertson *et al.*, 1996) some potyviruses (Revers *et al.*, 1996) and influenza orthomyxovirus (Webster *et al.*, 1995).

Phenotypes

Recently much research effort has been made to discover the molecular basis of virus host ranges. Much genetic and sequence information has accumulated to show that a wide range of positive and negative molecular traits, of both virus and host plant, are involved in the susceptibility and resistance of particular species (Dawson and Hilf, 1992). However the link between these observations and what is known of the host ranges of individual viruses, and of virus genera, is not clear. Much of the experimental evidence suggests that relatively trivial changes often have large effects on experimental host ranges, so it is surprising, therefore, that the major features of the host range of a virus seem to be relatively stable.

Virus isolates of a single species, with distinct symptoms, are usually found to differ only by a limited number of point mutations. Thus symptom expression is modulated by minor differences. This was probably best demonstrated when passaging through different host species was found to select variants that caused distinct symptoms (Yarwood, 1979). Experiments have also shown that single point mutations can reduce or eliminate the ability of a virus to infect a host species (Broglio, 1995). Single point mutations may also allow some virus variants to overcome host resistance mechanisms (Kavanagh *et al.*, 1992). However, experiments using strains or related virus species with distinct host ranges suggest that slightly more complex genetic change is usually required for host ranges to be modified experimentally. In one study, a single point mutation in a gene involved in replication (the alpha–a gene) allowed a barley infecting strain of barley stripe mosaic hordeivirus to infect oats, but a further five mutations in the gene were required for the level of replication of the virus to equal that of a strain of the virus that naturally infected oats (Weiland and Edwards, 1996). In another study, cowpea chlorotic mottle bromovirus carrying the 3a movement protein of brome mosaic bromovirus (BMV) would not infect cowpea plants unless a single codon change was also made in the gene. However, when this same change was made in the genome of BMV, it could not infect cowpea (Fujita *et al.*, 1996) suggesting that other genes may have a role.

Like host range, natural transmission may involve more than just one gene, and natural vector specificities are unlikely to be changed by a few simple point mutations. Seed transmission may involve several viral genes (Johansen, 1996), and whereas, for many viruses, the virion proteins and, for some, helper components, are the only points of direct contact with a vector, vector feeding preferences and the distribution of virions in a plant may also determine the success of transmission.

The genetic complexity of host range and transmission determinants may mean that these important virus characters only change when a series of mutations have accumulated. Recombination can short–circuit the need for a series of mutations by permitting the complete replacement of some genes with equivalent genes from another virus.

Predictions and possibilities : risks that may be caused by transgenic plants

Many laboratories around the world are currently investing much effort in producing crop plants that have been genetically transformed with viral genes to render them insusceptible to infection by the same or closely related viruses. These projects are based on the principle of "cross–protection" or "parasite–derived resistance" (McKinney 1929; Hamilton, 1980 ; Gibbs, 1982; Sanford and Johnston, 1985). Hopefully most of these transgenic plants will have the protection that is sought, and that this can be exploited quite safely. However, as this is an area of new and untried technology, caution is obviously important, and the potential hazards should be investigated.

In addition to possible risks posed by outcrossing and weediness (Cooper, 1997), transgenic plants carrying viral transgenes could affect existing virus populations in at least three deleterious ways:

1. Virulence may be modified when viruses are passaged through unusual hosts (Yarwood, 1979). Transgenic plants may, like any such host, select variants from the virus population, that are usually rare but are able to overcome the engineered resistance, and cause a greater problem than the original virus population;

2. The viral transgenes may genetically complement the activities of non–target viruses to cause new diseases (Candelier–Harvey and Hull, 1993; Lecoq *et al.*, 1993; Vance *et al.*, 1995; Ding *et al.*, 1997);

3. Recombination between viral transgenes and viruses still able to infect a transgenic plant, may generate a novel plant virus and a new disease (de Zoeten, 1991; Falk and Bruening, 1994).

The first of these possibilities will, most likely, produce variants with altered virulence. The second might produce altered viral phenotypes but not altered genotypes, and will not be maintained when the transgenic plant dies. However the third, recombination, might produce viruses with novel genotypes, and is most concerning as recent experiments have shown it to be real; viruses can acquire viral transgenes by recombination (Greene and Allison, 1994), viral recombination may increase the level of disease (Ding *et al.*, 1996), may allow viruses to infect new hosts (Schoelz and Wintermantel, 1993), and may produce 'fitter' viruses (Fernandez–Cuartero *et al.*, 1994 ; Wintermantel and Schoelz, 1996).

However several uncertainties have prevented a clear assessment of this risk. Firstly little is known about natural viral recombination and the factors that probably limit its frequency (AIBS, 1995). There may be critical barriers to natural recombination that will be overcome by the introduction of the viral transgene (Gibbs, 1994; Revers *et al.*, 1996; Allison *et al.*, 1997). Furthermore little is known about the factors affecting the survival of new recombinant viruses. It is well known that there are changes over time in the incidence and severity of viral crop diseases, but the part played by viral recombination in these changes is not known. Finally, although experiments have measured the frequency of intra–species recombination (Nagy and Bujarski, 1996; Carpenter and Simon, 1996), recombination between virus species from different genera and families has not yet been experimentally produced, or its frequency measured, and this suggests that it occurs infrequently.

Profiles for detecting phylogenetic inconsistencies

Recombination is usually detected by finding that different parts of a set of aligned genes have different relationships. Fig. 2 shows, for example, neighbor–joining trees (Saitou and Nei, 1987) representing the relationships of nine tobamoviruses; the tree on the left represents the relationships of aligned nucleotides 1–5175, namely the 5' untranslated and replicase regions, that on the right nucleotides 5176–7036, namely the movement protein, virion protein and 3' untranslated regions. Odontoglossum ringspot virus (ORSV) clearly shows a shift in its affiliations, as previously reported (Ryu and Park, 1995; Lartey *et al.*, 1996); in the 5' region it is most closely related to the tobamoviruses isolated from brassicas, whereas in the 3' region it is closest to those isolated from solanaceous plants. This phylogenetic inconsistency, which might have been caused by differences in evolutionary rates rather than recombination, could also have been detected by statistical techniques (Grassby and Holmes, 1997). However the human ability to recognise and compare patterns in graphically displayed data is probably more effective than purely numerical methods when studying complex data, such as nucleotide sequences. Therefore we have been designing graphical methods for detecting recombination, and have used these to investigate the tobamovirus data.

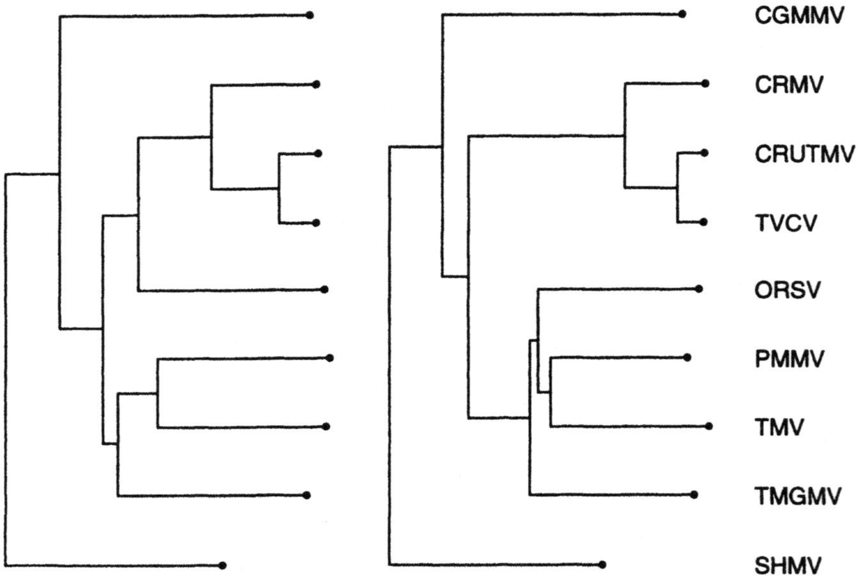

Fig. 2. Trees calculated from comparisons of tobamovirus sequences using the neighbor–joining method. The left tree was from comparisons of sequences within the RP gene (positions 1300–2300 in the condensed dataset, see text). The right tree was from comparisons of sequences within the MP and CP genes (positions 3500 to 4300 in the condensed dataset). Terminal nodes are labelled with virus acronyms as follows: odontoglossum ringspot tobamovirus (ORSV), tobacco mosaic tobamovirus (TMV), Chinese rape mosaic tobamovirus (CRMV), and sunnhemp mosaic tobamovirus (SHMV). The branching pattern of the two trees is identical except for the position of the terminal branch leading to ORSV.

Fig. 3 shows two 'phylogenetic profiles' of the aligned tobamovirus sequences obtained from the PhylPro program (Weiller, 1996). This program compares the relationships of aligned sequences within a split window that scans the sequences. At each position during the scan, each sequence is compared with every other sequence in both parts of the window to give two sets of distances.The correlation between these distances is the 'phylogenetic correlation', which ranges from +1.0 (perfectly correlated) to 0 (unrelated). Each sequence is compared, in turn, with every other sequence, and the phylogenetic correlations for all positions in the scan plotted in a single graph, the 'phylogenetic profile', so that the subtleties of their relationships can be seen. Two types of split window are used. In the first, the split is moved over the aligned sequences, and distances calculated from ALL the sequence on either side of the split (Fig. 3–A), whereas, in the second, a split window of fixed size is used to scan the sequences (Fig. 3–B) and thus the terminal portions of the sequences are within half of the window, and are lost.

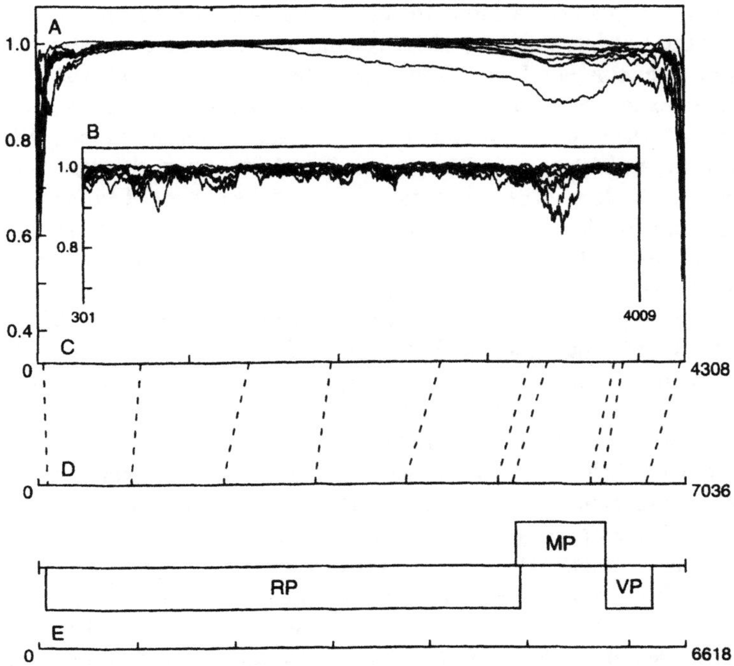

Fig. 3. A PhylPro plot showing the 'phylogenetic correlation profiles' of the genomic sequences of nine tobamoviruses; Chinese rape mosaic, crucifer TMV, cucumber green mottle mosaic, odontoglossum ringspot, pepper mild mottle, sunnhemp mosaic, tobacco mild green mosaic and tobacco mosaic tobamoviruses. Sequences were aligned, degapped and invariant positions removed. Phylogenetic correlations were calculated for all sequences on either side of a split–point, which scanned the entire alignment (uppermost profile–A) and for sequences within two adjacent windows 300 positions long (lower profile–B). The genome map of ORSV is shown with its size (scale E), together with the fully aligned and reassembled size (scale D), and the size after degapping and removing invariates (scale C); dotted lines join equivalent points. The recombination signal is clearest around nucleotide 3400 in the condensed alignments, namely just to the 5'–side of the junction of the RP and MP genes.

The PhylPro analysis of the tobamovirus genomic sequences shown in Fig. 3 proceeded as follows. The sequences were first split into individual genes (including overlapping portions), translated into their encoded amino acid sequences, aligned using ClustalV (with default parameters), and then triplets of gaps added to appropriate places in the gene sequences, which were then rejoined to give aligned genomes 7036 nucleotides long (Fig. 3–D); that of ORSV started 6618 nucleotides long (Fig. 3–E). All invariant positions and gaps were then removed resulting in sequences 4308 nucleotides long (Fig. 3–C) for PhylPro analyses. The figure also shows the relative positions in the original sequences, then the gapped aligned sequences, and finally the 'gapped and informative' sequences; the nucleotides every 1000 nucleotides from the 5' terminus, as well as at gene boundaries, are marked. The upper profile (full sequence, moving boundary) clearly shows that one sequence, that of ORSV, has a broad area of phylogenetic inconsistency at the RP/MP boundary, and the position of this is confirmed and shown more precisely in the lower profile (fixed 300:300 window).

However it requires a more focussed approach to confirm that ORSV is a recombinant, not just that part of its genome has evolved in an atypical manner. Therefore we compared the ORSV sequence with those closest to it in its alternative taxonomic positions, namely tobacco mosaic virus (TMV) and Chinese rape mosaic virus (CRMV), and that of sunnhemp mosaic virus (SHMV) was included as an outgroup. The 'gapped and informative' aligned nucleotide sequences of the four viruses were examined and the pattern of support throughout the alignment for the relationships of the four was determined using the program SplitsPro. At each position there are thirteen possible patterns of the four nucleotides (the invariant positions had been removed) that are generally informative; sequences 1 and 2 the same, and both of the others different, next, sequences 1 and 3 the same and both of the others different, etc, etc. The pattern was assessed at each position in the alignment, and the support throughout the sequence for four of the patterns presented as a graph (Fig. 4). It can be seen that, to the 5'-side of nucleotide 3200, and despite much variation, there is more support for an ORSV/CRMV grouping, than an ORSV/TMV grouping (70 positions support each of patterns containing an ORSV/CRMV grouping compared with an average of 39.4 supporting each the other patterns), whereas the support is reversed to the 3'-side of nucleotide 3200. To check whether the support for these patterns merely resulted from biased nucleotide composition in these regions, we also shuffled the nucleotides within positions in the regions bounded by nucleotides 1300–2300, and by 3500–4300, and determined support for the thirteen patterns in the randomised data one hundred times; these regions were chosen as the region immediately 5' of the RP/MP boundary (around nucleotide 3200) has confused pattern support. The pattern support for the randomised data was used as 'expected values', and gave 'pseudo Chi^2' values of 152.5 and 153.5 respectively, indicating that the probability that support for the groupings had arisen by chance within each region was very considerably less than 0.001.

ORSV infects orchids, whereas the two most closely related sets of tobamoviruses mostly infect brassicas and solanaceous plants respectively. The

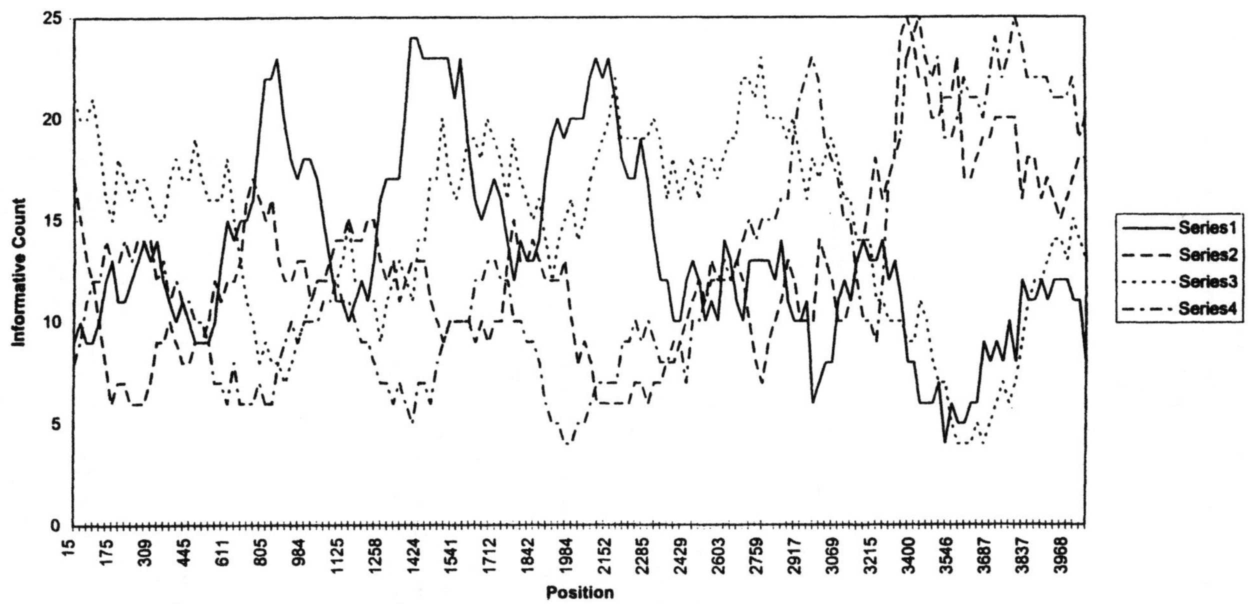

Fig. 4. A SplitsPro plot showing the distribution of four of the thirteen kinds of generally informative nucleotide patterns in the 'condensed' aligned genome sequences of ORSV, TMV, CRMV, and SHMV. The numbers of each type of pattern were counted in a window 100 positions wide, and the window moved through the alignment in 10–position steps. Series 1 sites are those where ORSV and CRMV have identical nucleotides and the others different, series 2 are where ORSV and TMV have identical nucleotides and the others different, series 3 are where ORSV and CRMV are the same and SHMV and TMV are the same, and, finally, series 4 sites are where ORSV and TMV are the same and CRMV and SHMV are the same.

ability of ORSV to infect orchids has been shown to depend minimally on some function of its movement protein (Fenczik *et al.*, 1995). It is, of course, most likely that the recombination that produced the ancestral ORSV preadapted it to survive and prosper in orchids, however there is also the possibility that the recombination event and the host change were not causally related, indeed it may not now be possible to unravel the events, because of subsequent pleiotropic genomic adaptations. Nonetheless the possible link indicates the value of investigating other viruses that have anomalous hosts to see whether they too are recombinants.

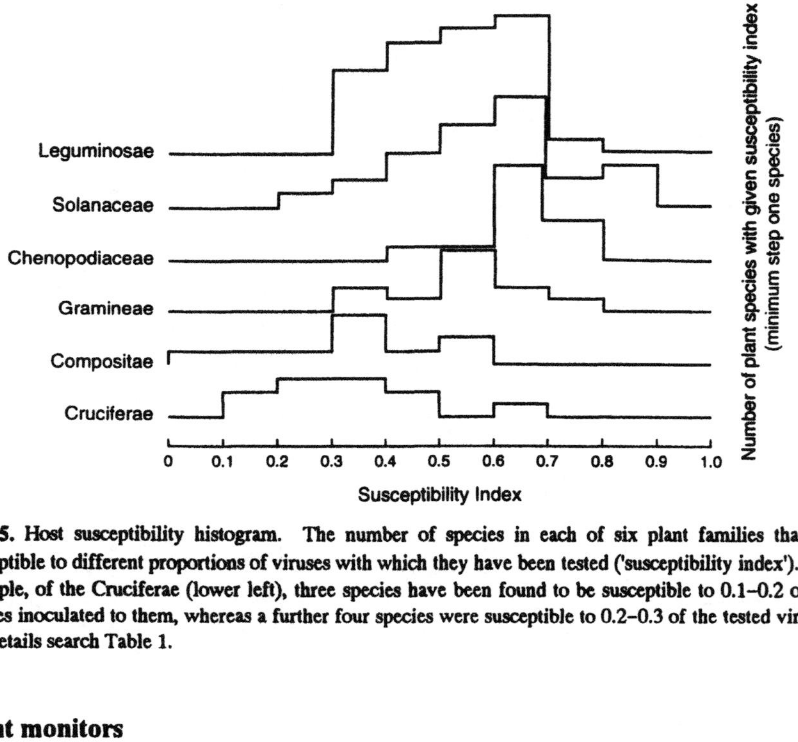

Fig. 5. Host susceptibility histogram. The number of species in each of six plant families that are susceptible to different proportions of viruses with which they have been tested ('susceptibility index'). For example, of the Cruciferae (lower left), three species have been found to be susceptible to 0.1–0.2 of the viruses inoculated to them, whereas a further four species were susceptible to 0.2–0.3 of the tested viruses. For details search Table 1.

Plant monitors

A major genetic change in a plant virus genome, such as that resulting from recombination with a transgene, is unlikely merely to change its virulence in its customary hosts. One possibility is that a variant with a quite different host range will be produced, thus responsible monitoring of transgenic crops during their development should involve strategies that would detect variants with altered host ranges. Choice of suitable bait plant species for this purpose is crucial. The VIDE database (Brunt *et al.*, 1996a and 1996b) provides some help. It contains 18,243 virus/plant species records, involving 1682 plant species (141 families) and 912 viruses (56 genera), and ranging from the single virus records for 961 plant species to the 471 virus species inoculated to *Chenopodium amaranticolor*. Although it is the most complete record of its sort, it is clearly very incomplete and biased, most obviously as a result of the fact that many published reports of host data omit records of immune species.

The VIDE records show a few patterns. For example Monte Carlo analyses (J. Armstrong and A.J. Gibbs, unpublished) show that about one fifth of the viruses infect related plants (judged from similarities of their RuBisCo sequences, or of their phenotypic properties; Watson and Dallwitz, 1992) more frequently than unrelated ones. Also obvious is that viruses first isolated from monocotyledonous plants infect more species of that group, than dicotyledonous plants, however viruses that clearly 'prefer' related plants are scattered throughout most large viral genera. Plant susceptibilities also show some patterns. Fig. 5 shows, for example, the number of species from particular plant families that have proven susceptible to different proportions of the viruses inoculated to them (i.e. 'susceptibility indices'). It can be seen that brassicas and composites are, on average, only infected by half as many viruses as chenopods and others. Table 1 lists the plant species for which there are 25 or more records in the VIDE database, and gives the proportion of viruses to which they are known to be susceptible. There are no clear patterns linking the susceptibility indices of different plant species and the viral genera to which they are susceptible. However we suggest that all novel viruses, whether recombinant or not, are most likely to infect plant species with a large susceptibility indices, indeed one of the more susceptible, *Chenopodium hybridum*, was found by Rao *et al.* (1990) to "facilitate isolation of brome mosaic virus RNA recombinants". Although the value of particular plant species as viral 'bait' may be modulated by the host preferences of vectors. Plants with intermediate susceptibility indices are clearly best for use as indicator plants.

Table 1. The species of plants for which there are susceptibility records for 25 or more viruses; A. number of viruses tested; B. proportion of viruses that infected

Host species	A	B	Host species	A	B
Dahlia pinnata	33	0.09	*Matthiola incana*	31	0.45
Chrysanthemum morifolium	25	0.16	*Citrullus lanatus*	60	0.45
Brassica oleracea var. capitata	68	0.16	*Vigna unguiculata ssp. sesquipedalis*	76	0.45
Verbesina encelioides	26	0.19	*Beta vulgaris*	227	0.45
Brassica campestris ssp. pekinensis	108	0.19	*Cassia tora*	37	0.46
Raphanus sativus	96	0.20	*Petroselinum crispum*	39	0.46
Brassica oleracea var. botrytis	61	0.26	*Cajanus cajan*	50	0.46

Host species	A	B	Host species	A	B
Brassica campestris ssp. rapa	92	0.28	*Antirrhinum majus*	89	0.46
Capsicum annuum	178	0.28	*Catharanthus roseus*	110	0.46
Ageratum conyzoides	28	0.29	*Phaseolus vulgaris*	442	0.46
Gossypium hirsutum	51	0.29	*Phlox drummondii*	43	0.47
Allium cepa	33	0.30	*Oryza sativa*	48	0.48
Abelmoschus esculentus	61	0.31	*Sonchus oleraceus*	43	0.49
Medicago sativa	105	0.31	*Phaseolus lunatus*	75	0.49
Lactuca sativa	169	0.31	*Nicotiana glutinosa*	407	0.49
Dactylis glomerata	34	0.32	*Lolium perenne*	36	0.50
Cheiranthus cheiri	30	0.33	*Trifolium hybridum*	48	0.50
Fagopyrum esculentum	33	0.33	*Triticum aestivum*	107	0.50
Brassica juncea	41	0.34	*Physalis floridana*	176	0.50
Dianthus barbatus	67	0.34	*Avena sativa*	97	0.51
Trifolium pratense	105	0.34	*Nicotiana tabacum*	451	0.51
Solanum melongena	128	0.34	*Melilotus albus*	67	0.52
Brassica campestris	26	0.35	*Datura metel*	95	0.52
Pastinaca sativa	37	0.35	*Petunia x hybrida*	306	0.52
Lablab purpureus	49	0.35	*Cucurbita moschata*	49	0.53
Calendula officinalis	54	0.35	*Vigna radiata*	86	0.53
Helianthus annuus	102	0.35	*Hordeum vulgare*	106	0.53
Cucumis sativus	408	0.35	*Nicotiana rustica*	216	0.53
Arachis hypogaea	109	0.36	*Vigna angularis*	26	0.54
Trifolium repens	123	0.37	*Sesamum indicum*	48	0.54
Zea mays	154	0.37	*Solanum nigrum*	85	0.54
Vigna unguiculata ssp. cylindrica	45	0.38	*Amaranthus caudatus*	92	0.54
Zinnia elegans	201	0.38	*Glycine max*	201	0.54
Capsicum frutescens	96	0.39	*Senecio vulgaris*	40	0.55
Cucurbita pepo	159	0.39	*Trifolium incarnatum*	125	0.55
Momordica balsamina	30	0.40	*Sesbania exaltata*	27	0.56

Host species	A	B	Host species	A	B
Brassica campestris ssp. napus	35	0.40	*Chenopodium hybridum*	34	0.56
Daucus carota	82	0.40	*Celosia cristata*	30	0.57
Dianthus caryophyllus	44	0.41	*Secale cereale*	51	0.57
Vicia faba	261	0.41	*Gomphrena globosa*	393	0.57
Apium graveolens	89	0.42	*Cichorium endiva*	33	0.58
Cucurbita maxima	98	0.42	*Amaranthus retroflexus*	43	0.58
Cucumis melo	108	0.42	*Cassia occidentalis*	60	0.58
Solanum tuberosum	130	0.42	*Celosia argentea*	66	0.58
Lycopersicon esculentum	321	0.42	*Crotalaria spectabilis*	49	0.59
Tropaeolum majus	65	0.43	*Phytolacca americana*	25	0.60
Ocimum basilicum	72	0.43	*Solanum demissum*	30	0.60
Pisum sativum	288	0.43	*Trifolium subterraneum*	42	0.60
Vigna unguiculata	325	0.44	*Cyamopsis tetragonoloba*	90	0.60
Datura stramonium	357	0.44	*Spinacia oleracea*	181	0.60
Vicia villosa	28	0.61	*Nicotiana debneyi*	142	0.67
Lycopersicon pimpinellifolium	41	0.61	*Tetragonia tetragonioides*	198	0.67
Cicer arietinum	46	0.61	*Chenopodium ambrosioides*	31	0.68
Lupinus albus	54	0.61	*Nicotiana x edwardsonii*	36	0.69
Sorghum bicolor	29	0.62	*Torenia fournieri*	36	0.69
Coriandrum sativum	42	0.62	*Chenopodium foetidum*	68	0.69
Hyoscyamus niger	54	0.63	*Lathyrus odoratus*	83	0.69
Nicotiana sylvestris	90	0.63	*Chenopodium album*	144	0.69
Atriplex hortensis	47	0.64	*Stellaria media*	53	0.70
Capsella bursa-pastoris	47	0.64	*Chenopodium murale*	192	0.70

Host species	A	B	Host species	A	B
Chenopodium capitatum	81	0.64	*Nicotiana clevelandii*	377	0.70
Nicandra physalodes	109	0.64	*Chenopodium foliosum*	32	0.72
Fragaria vesca	29	0.66	*Chenopodium quinoa*	464	0.72
Lupinus angustifolius	29	0.66	*Nicotiana megalosiphon*	128	0.73
Macrotyloma uniflorum	29	0.66	*Vicia sativa*	42	0.74
Lens culinaris	41	0.66	*Lolium multiflorum*	31	0.77
Physalis peruviana	53	0.66	*Nicotiana occidentalis*	41	0.80
Chenopodium amaranticolor	471	0.66	*Nicotiana benthamiana*	201	0.80
Setaria italica	27	0.67	*Nicotiana bigelovii*	40	0.83
Macroptilium lathyroides	42	0.67			

Postscript

Our knowledge of the processes of viral evolution is based on fragmentary gene sequence evidence from a relatively small number of viruses, most of them in epidemic mode. Molecular experiments on the most significant biological features of viruses, such as their host ranges preferences, virulence, 'fitness' and vector interactions, have only just begun, and are mostly reductionist in principle. Information from these two sources is not completely congruent. The discovery that recombination is a significant feature of potyvirus evolution (Revers *et al.*, 1996), contrary to recently held views (APHIS, 1995), and that type I and II strains of cucumber mosaic cucumovirus may be speciating (Fraile, 1997), emphasizes the uncertainty. Under these circumstances, only the foolhardy will have firm views on the likely outcome of the introduction of virus–resistant transgenic crops into agriculture. Much more information is required about the evolution and survival of natural populations of viruses in weeds, wild plants and crop plants, before there can be greater certainty.

Acknowledgement

We thank Eric J. Zurcher for extracting the data in Table 1 from the VIDE database.

References

AIBS (1995) Proceedings of the workshop, Transgenic virus-resistant plants and new plant viruses. 20-21 April 1995, Beltsville, Maryland, American Institute of Biological Sciences. 47 pp. Internet address: http://www.reston.com/ aibs/transgenic/virus.html

APHIS (1995) Petition Determination, number 9220401. Animal and Plant Health Inspection Service. Internet address: ftp://www.aphis.usda.gov/pub/bbep/ 9220401P.DET

Allison RF, Greene AE, Schneider WL (1997) Considering the risks of virus resistant transgenic crops. In: McLean GD , Waterhouse PM, Evans GA, Gibbs MJ (ed). Commercialisation of Transgenic Crops: Risk Benefit and Trade Considerations. CRC for Plant Science and Bureau of Resource Sciences, Canberra. pp. 145

Baulcombe DC (1996) Mechanisms of pathogen-derived resistance to viruses in transgenic plants. Plant Cell 8:1833-1844

Broglio EP (1995) Mutational analysis of cauliflower mosaic virus gene VI: Changes in host range, symptoms, and discovery of transactivation-positive, noninfectious mutants. Molec Plant-Microbe Interact 8:755-760

Brunt A, Crabtree K, Dallwitz M, Gibbs A, Watson L (1996a) Viruses of Plants: Descriptions and Lists from the VIDE Database. CAB International, Wallingford, United Kingdom. 1484 pp

Brunt A, Crabtree K, Dallwitz M, Gibbs A, Watson L, Zurcher EJ (1996b) Plant Viruses Online. Internet address. http://biology.anu.edu.au/research-groups/MES/vide/

Candelier-Harvey P, Hull R (1993) Cucumber mosaic virus genome is encapsidated in alfalfa mosaic virus coat protein expressed in transgenic tobacco plants. Transgenic Res 2:277-285

Carpenter CD, Simon AE (1996) In vivo restoration of biologically active 3' ends of virus-associated RNAs by nonhomologous RNA recombination and replacement of a terminal motif. J Virol 70:478-486

Cooper JI (1997) Might transgenes conferring virus resistance harm the environment? In: McLean GD , Waterhouse PM, Evans GA, Gibbs MJ (ed). Commercialisation of Transgenic Crops: Risk Benefit and Trade Considerations. CRC for Plant Science and Bureau of Resource Sciences, Canberra. pp. 115-123

Deom CM, He ZZ, Beachy RN, Weissinger AK (1994) Influence of heterologous tobamovirus movement protein and chimeric-movement protein genes on cell-to-cell and long-distance movement. Virology 205:198-209

de Zoeten G (1991) Risk assessment: Do we let history repeat itself? Phytopathology 81:585-586

Ding SW, Shi BJ, Li WX, Symons RH (1996) An interspecies hybrid RNA virus significantly more virulent than either parental virus. Proc Natl Acad Sci USA 93:7470-7474

Ding SW, Afsharifar A, Shi BJ Li WX, Symons RH (1997) Recombinant viruses exhibiting a nonconventional type of virus synergy: The relevance to risk assessment of transgenic virus-resistant crops. In: McLean GD , Waterhouse PM, Evans GA, Gibbs MJ (ed) Commercialisation of Transgenic Crops: Risk Benefit and Trade Considerations. CRC for Plant Science and Bureau of Resource Sciences, Canberra. pp. 151-158

Domingo E, Holland JJ (1994) Mutation rates and rapid evolution of RNA viruses. In: Morse SS (ed) The Evolutionary Biology of Viruses. Raven Press, New York, pp. 161-18

Domingo E, Sabo D, Taniguchi T, Weissmann C (1978) Nucleotide sequence heterogeneity of an RNA phage population. Cell 13:735-744

Edwards MC, Zhang Z, Weiland JJ (1997) Oat blue dwarf marafivirus resembles the tymoviruses in sequence, genome organization and expression strategy. Virology, in press

Fenczik CA, Padgett HS, Holt CA, Casper SJ, Beachy RN (1992) Mutational analysis of the movement protein of odontoglossum ringspot virus to identify a host-range determinant. Molec Plant-Microbe Interact 8:666-673

Fernandez-Cuartero B, Burgyan J, Aranda MA, Salanki K, Moriones E, Garcia-Arenal F (1994) Increase in the relative fitness of a plant virus RNA associated with its recombinant nature. Virology 203:373-377

Fraile A, Alonso-Prados JL Aranda MA Bernal JJ, Malpica JM, Garcia-Arenal F (1997) Genetic exchange by recombination or reassortment is infrequent in natural populations of a tripartite RNA plant virus. J Virol 71:934-940

Gibbs A (1982) Competition between viruses; a transposable phenomenon? Genetic Engineering for Agriculture. Occasional Publication No. 2 of the Australian Institute of Agricultural Science

Gibbs A. (1987) Molecular evolution of viruses; 'trees', 'clocks' and 'modules'. J Cell Sci, Suppl 7:319-337

Gibbs A, Keese, P. (1995) In search of the origins of viral genes. In: Gibbs AJ, Calisher CH, Garcia-Arenal F (ed) Molecular Basis of Virus Evolution. Cambridge University Press, Cambridge. pp. 76-90

Gibbs MJ (1994) Risks in using transgenic plants? Science 264:1650-1651

Gibbs MJ, Cooper JI (1995) A recombinational event in the history of luteoviruses probably induced by base-pairing between the genomes of two distinct viruses. Virology 206:1129-1132

Gibbs MJ, Cooper JI, Waterhouse PM (1996) The genome organisation and affinities of an Australian isolate of carrot mottle umbravirus. Virology 224:310-313

Goldbach R (1992) The recombinative nature of potyviruses: Implications for setting up a true phylogenetic taxonomy. Arch Virol, Suppl 5:299-304

Gorbaleny AE (1992) Host-related sequences in RNA viral genomes. Semin Virol 3:359-371

Grassby NC, Holmes EC (1997) A likelihood method for the detection of selection and recombination using nucleotide sequences. Molec Biol Evol 14:239-247

Greene AE, Allison RF (1994) Recombination between viral RNA and transgenic plant transcripts. Science 763:1423-1425

Hamilton RI (1980) Defenses triggered by previous invaders: viruses. In: Horsfall JG, Cowling EB (ed) Plant Disease, An Advanced Treatise, vol. 5. pp 279-303 Academic Press, New York

Hancock JM, Chaleeprom W, Chaleeprom W, Dale J, Gibbs A (1995) Replication slippage in the evolution of potyviruses. J Gen Virol 76:3229-323

Johansen IE, Dougherty WG, Keller KE, Wang D, Hampton RO (1996) Multiple viral determinants affect seed transmission of pea seedborne mosaic virus in Pisum sativum. J Gen Virol 77:3149-3154

Johnson GP, Goebel SJ, Paoletti E (1993) An update on the vaccinia virus genome. Virology 196:381-401

Kamer G, Argos P (1984) Primary structural comparison of RNA-dependent polymerases from plant, animal and bacterial viruses. Nucl Acids Res 12:7269-7282

Kavanagh T, Goulden M, Santa Cruz S, Chapme S, Barker I, Baulcombe D (1992) Molecular analysis of a resistance-breaking strain of potato virus X. Virology 189:609-617

Keese PK, Gibbs AJ (1992) Origins of genes: "Big bang" or continuous creation? Proc Natl Acad Sci USA. 89:9489-9493

Koonin EV, Dolja VV (1993) Evolution and taxonomy of positive-strand RNA viruses: implications of comparative analysis of amino acid sequences. Crit Rev Biochem Molec Biol 28:375-430

Lartey RT, Voss TC, Melcher U (1996) Tobamovirus evolution: gene overlaps, recombination and taxonomic implications. Molec Biol Evol 13:1327-133

Lecoq H, Ravelonandro M, Wipf-Scheibel C, Monsion M, Raccah B, Dunez J (1993) Aphid transmission of a non-aphid transmissible strain of zucchini yellow mosaic potyvirus from transgenic plants expressing the capsid protein of plum pox potyvirus. Molec Plant-Microbe Interact 6:403-406

McKinney HH (1929) Mosaic diseases in the Canary Islands, West Africa and Gibraltar. J Agric Res 39:557-578

Nagy PD, Bujarski JJ (1996) Homologous RNA recombination in brome mosaic virus AU-rich sequences decreases the accuracy of crossovers. J Virol 70:415-426

Rao AL, Sullivan BP, Hall TC (1990) Use of Chenopodium hybridum facilitates isolation of brome mosaic virus RNA recombinants. J Gen Virology 71:1403-1407

Rathjen JP, Karageorgos LE, Habili N, Waterhouse PM, Symons RH (1994) Soybean dwarf luteovirus contains the third variant genome type in the luteovirus group. Virology 198:671–679

Revers F, Legall O, Candresse T, Leromancer M, Dunez J (1996) Frequent occurrence of recombinant potyvirus isolates. J Gen Virol 77:1953–1965

Robertson DL, Hahn BH, Sharp PM (1995) Recombination in AIDS viruses. J Molec Evol 40:249–259

Ryu KH, Park WM (1995) The complete nucleotide sequence and genome organization of Odontoglossum ringspot tobamovirus RNA. Arch Virol 140:1577–1587

Saitou N, Nei M (1987) The neighbour-joining method: A new method for reconstructing phylogenetic trees. Molec Biol Evol 4:406–425

Sanford JC, Johnston SA (1985) The concept of parasite derived resistance – deriving resistance genes from the parasite's own genome. J Theor Biol 113:395–405

Schoelz JE, Wintermantel WM (1993) Expansion of viral host range through complementation and recombination in transgenic plants. Plant Cell 5:1669–1679

Steinhauer DD, Domingo E, Holland JJ (1992) Lack of evidence of proofreading mechanisms associated with an RNA virus polymerase. Gene 122:281–288

Vance VB, Berger PH, Carrington JC, Hunt AG, Shi XM (1995) 5' proximal potyviral sequences mediate potato virus X/potyviral synergistic disease in transgenic plants Virology 206:583–590

Veidt I, Lot H, Leiser M, Scheidecker D, Guilley H, Richards K, Jonard G (1988) Nucleotide sequence of beet western yellows virus RNA. Nucl Acids Res 16:9917–9932

Watson L, Dallwitz MJ (1992) The families of angiosperms. Internet address: http://huh.harvard.edu/pub/delta/angio

Weaver SC (1995) Evolution of alphaviruses. In: Gibbs AJ, Calisher CH, Garcia-Arenal F (ed) Molecular Basis of Virus Evolution. Cambridge University Press, Cambridge. pp. 501–530

Webster RG, Bean WJ, Gorman OT (1995) Evolution of influenza viruses: rapid evolution and stasis. In: Gibbs AJ, Calisher CH, Garcia-Arenal F (ed) Molecular Basis of Virus Evolution. Cambridge University Press, Cambridge. pp. 531–543

Weiland JJ, Edwards MC (1996) A single nucleotide substitution in the alpha-a gene confers oat pathogenicity to barley stripe mosaic virus strain ND18. Molec Plant-Microbe Interact 9:62–67

Weiller GF (1996) Phylogenetic profiles: A graphical method for detecting recombination in homologous sequences Molec Biol Evol (submitted)

Wintermantel WM, Schoelz JE (1996) Isolation of recombinant viruses between cauliflower mosaic virus and a viral gene in transgenic plants under conditions of moderate selection pressure. Virology 223: 156–164

Yarwood CE (1979) Host passage effects with plant viruses. Adv Virus Res 25:169–190

Zanotto PM de A, Gould EA, Gao, GF, Harvey PH, Holmes EC (1996) Population dynamics of flaviviruses revealed by molecular phylogenies. Proc Natl Acad Sci USA 93:546–553

Systematic Search for Recombination Events in plant Viruses and Viroids

Thierry Candresse[1], Frédéric Revers[1], Olivier Le Gall[1], Sandra A. Kofalvi[1], José Marcos[2] and Vicente Pallás[2]
[1] Station de Pathologie Végétale, INRA, BP 81, 33883 Villenave d'Ornon Cedex, France
[2] Centro de Edafología y Biología del Segura, CSIC, Av. de la Fama 1, 33003 Murcia, Spain

Abstract

One of the key issues regarding the biosafety of transgenic plants expressing plant virus genetic information concerns the potential for recombination between transgene-derived sequences and an incoming, more or less related virus. With no precise information currently available on the rate of appearance and potential fitness of such recombinants, the hypothesis that all viable possible recombinants have already been produced and selected in nature in the course of natural multiple infections has been the subject of a debate. One way to further our understanding of these mechanisms is to perform systematic searches, in order to identify natural viral isolates deriving from a recombination event and, therefore, make a first evaluation of the frequency of recombinant isolates in natural viral populations.

We have performed such a systematic search using the coat protein gene (CP gene) and 3' non-coding regions of all potyviruses for which more than 4 independent sequences were available in the databanks. In the case of potato virus Y (PVY), close to one third of the isolates present in the databanks derived from a recombination event, identifying five different breakpoints scattered over the region analysed. In the case of bean common mosaic virus (BCMV) two isolates out of 15 analysed proved to derive from a recombination event. On the contrary, and despite large datasets, no recombination events could be detected in the region analysed for several other viruses such as plum pox virus (PPV) or papaya ringspot virus (PRSV). It therefore appears that potyviruses may differ in their intrinsic capacity to recombine or, alternatively, that for some potyviruses the potential for the generation of new isolates through recombination has not yet be fully explored.

We have recently performed similar systematic searches for two other pathogens for which large sequence datasets are available. In the case of cucumber mosaic cucumovirus (CMV) no evidence for recombination could be detected using complete sequences of genomic RNAs (respectively 6 RNA1, 12 RNA2 and 21 RNA3 sequences analysed). On the contrary, using a very different pathogen, hop stunt viroid (HSVd), evidence was obtained that a high proportion of HSVd isolates analysed resulted from recombination events.

These observations indicate that different viruses or viroids may present very different situations as far as the frequency of recombinant isolates in natural populations is concerned, and that if some viruses may already have reached their equlilibrium regarding this evolutionary driving force, such may not be the case for others.

Introduction

Although the risks involved may be the subject of very different analyses, one of the key issues regarding the biosafety of transgenic plants expressing plant virus genetic information concerns the potential for recombination between transgene-derived sequences and an incoming, more or less related, virus (Tepfer, 1993; Falk and Bruening 1994). There is now a great deal of evidence, both direct and indirect, that RNA viruses and viroids may exchange genetic material as a result of RNA recombination, both within and outside of their species. Similarly, recent evidence has demonstrated that both DNA and RNA viruses may acquire transgene-derived sequences through recombination (Gal *et al.*, 1992 ; Greene and Allison, 1994). There is therefore little doubt that recombinant viruses incorporating transgene-derived sequences will ultimately appear as a consequence of the widespread use of transgenic crops protected against viral infection by the expression of viral sequences. With no precise information currently available on the rate of appearance of such recombinants or on their potential fitness, the comforting hypothesis that all possible viable recombinants have already been produced in the course of natural multiple infections (and hence that no additional risk is involved with transgenics) has focused much of the debate.

It is generally thought that a recombinant must show some selective advantage over its parents in order to become established in nature and to overtake a complex viral population. This argument in itself disregards the potential impact of other evolutionary forces such as genetic bottlenecks (could vector transmission events be suitable ?). Indeed, experimental work has demonstrated that accumulation of genetic bottlenecks may lead to decreased fitness of viral populations and thus to the fixation of variants of lower fitness (Clarke *et al.*, 1993). As a means to explore the possibility of fixation of recombinant isolates of plant viruses and viroids in natural populations, we have performed a systematic search for recombinant isolates using nucleotide sequences deposited in databanks. The strategy used was to compare phylogenies reconstructed from different parts of the genome. Assuming coordinated evolution of all parts of the genome, all reconstructions should show similar topologies. Any shift in the affinities of a given isolate is thus likely to reflect the influence of recombination events. The recombinations thus detected were then validated using statistical analysis. A similar approach allowed Chenault & Melchers (1994) to demonstrate the frequent occurrence of recombinant isolates in a plant pararetrovirus, cauliflower mosaic virus.

Materials and methods

The sequences and strategy used to search for recombination events in the coat protein (CP) gene and in the 3' non-coding region of potyviruses has already been described (Revers *et al.*, 1996). In the case of cucumber mosaic cucumovirus (CMV) all full-length sequences of genomic RNAs were extracted from databanks and analysed : D12537, MCVFRNA1, MCVL1, CMU20220, MCV1A1 and CURNA1Q (RNA1); MCVRN2, MCVORNA2, CMU66287, CMU59740, MCVL2, D86613, S72187, MCV2A2, D86330, D12538, CMU20218 and CVRNA02 (RNA2); and MCVRNA3, MCVRNA3A, MCV3APCOAT, CMV3ACP, MCVCMVD, MCV3A3, MCVCMVA, MCVPT7CKY3,

MCVYRNA3, MCVR3MPCP1, MCV3ACP, MCVCP3A2, MCVCMVC, MCVR3MPCP2, MCVCMVB, MCVL3, MCVCP3A1, CMU66094, MCV03, MCVRNA3KOR and CMU20219 (RNA3). In the case of hop stunt viroid (HSVd), all the nucleotide sequences available in the recently created viroid web site (Bussière *et al.*, 1996) together with 12 other complete sequences (three which have been previously published and nine determined by our two groups (Kofalvi, Marcos *et al.*, submitted). A total of 32 full-length sequences of HSVd isolates were thus analysed.

For computer analysis of the sequences, the techniques described by Revers *et al.*, (1996) were used. Briefly, multiple alignments were obtained using the program ClustalV (Higgins *et al.*, 1992). Phylogenies were reconstructed and validated using the neighbour-joining technique as implemented in the PHYLIP package (Felsenstein, 1989). For statistical validation of the recombinations detected, the statistical test developed by Sawyer (1989) was used, as implemented in VTDIST2 (program provided by Dr. S. Sawyer). The recombination breakpoints were tentatively localised using the maximum chi-squared approach described by Maynard-Smith (1992), as implemented in the program RecSite (Le Gall, unpublished).

Results

A systematic search for recombination events in the coat protein (CP) gene and, when available, in the 3' non-coding region (3'NCR) of all potyviruses for which more than 4 independent CP sequences were available in the public databanks was performed (Revers *et al.*, 1996). Surprisingly, very different situations are observed from one virus to the other. In the case of potato virus Y (PVY), close to one third (11/31) of the sequences present in the databanks were found to be derived from recombination events, identifying six different breakpoints scattered over the region analysed. In the case of bean common mosaic virus (BCMV) two isolates out of 15 analysed proved to be derived from a recombination event. Recombination was also suspected on smaller numbers of sequences in the case of bean yellow mosaic virus (BYMV) and zucchini yellow mosaic virus (ZYMV). However, in these last two cases, recombination could not be statistically supported either because the isolates involved were too closely related or because one of the suspected parents was missing. On the contrary, and despite large datasets, no recombination events could be detected in the region analysed for several other viruses such as plum pox virus (PPV, 22 sequences), papaya ringspot virus (PRSV, 16 sequences), turnip mosaic virus (TuMV, 8 sequences) and watermelon mosaic virus 2 (WMV2, 4 sequences). It therefore appears that, at least for some potyviruses, recombinant isolates may constitute a very significant fraction of the viral population. The overall rate of recombinant isolates, observed or suspected in four of the eight potyviral species analysed, can be estimated to represent between 8 and 17% of the 109 isolates analysed, depending on the calculations (Revers *et al.*, 1996). These results, showing very contrasted situations between different potyviruses, also indicate either differences in the intrinsic capacity to recombine of these viruses or, alternatively, variations in their biological or epidemiological characteristics, so that the potential for the generation of new isolates through recombination has not yet been fully explored for some potyviruses.

In an effort to expand this approach, we have recently performed similar systematic searches for two other pathogens for which large sequence datasets are available. In the case of cucumber mosaic cucumovirus (CMV) (6 RNA1, 12 RNA2 and 21 RNA3 complete sequences analysed respectively) no evidence for recombination could be detected. On the contrary, using a very different pathogen, hop stunt viroid (HSVd), evidence was obtained that a high proportion of the 32 HSVd isolates analysed resulted from recombination events. Hop stunt viroid is able to infect a diverse host range including *Prunus* and *Citrus* species, grapevine, hop, cucumber and several other reported hosts. Earlier phylogenetic analyses had shown that three groups of isolates could be discriminated (Hsu *et al.*, 1994). Since these groups showed a clear tendency to contain isolates with identical isolation hosts, they were named plum–, citrus– and hop–type groups respectively. The present phylogenetic analysis on a larger dataset confirmed and extended these results by showing the existence, in addition to these three major groups, of two minor groups containing three grapevine variants and two *Prunus* variants respectively. Analysis of potential recombination events indicated that the isolates forming these two minor groups are likely to result from recombination events. In addition, the hop–type group of isolates itself is probably the result of recombination events between members of the plum–type and citrus–type groups. In total, this analysis indicated that 18 out of the 32 HSVd isolates studied could be considered to be recombinant isolates (Kofalvi *et al.*, submitted). Although interspecific recombination has long been known to be one of the driving forces behind viroid evolution (Keese and Symons, 1985; Hammond *et al.*, 1989), this is the first demonstration of its widespread effects in generating intraspecific variability.

Discussion and conclusion

These observations indicate that the use of the strategy described here for systematic searches of nucleic acids sequences present in databanks may allow the identification of recombinant isolates for both viruses and viroids. They also indicate that very different situations can be observed as far as the frequency of recombinant isolates in natural populations is concerned. Part of the differences observed from one virus to another may result from the limitations of this strategy. For example, it is very difficult to detect recombination events between closely related isolates (at the extreme, self–recombination is by essence undetectable!) or whenever one of the parental isolates is absent from the dataset, as demonstrated by the difficulties encountered with some potyviruses. Also, the uneven distribution of selection pressures along a viral genome, leading to the juxtaposition of highly conserved and variable regions or the presence of highly variable isolates in a dataset may exert confounding effects. For example, direct analysis of the CMV dataset using the VTDIST2 programs gives values highly significant in favour of recombination. Careful analysis reveals, however, that these results are linked to the presence of two highly divergent subgroups of CMV isolates and of regions of high sequence conservation such as the non–coding regions.

One interesting point is that further experimental work has, to some extent, validated the results presented here. For example, PCR–RFLP analysis of PVY isolates has led Kerlan *et al.* (1997) to conclude that recombinant PVY isolates do exist in field samples, while a thorough analysis of over 250 field isolates of CMV conducted by Fraile *et al.*, (1997) has shown that recombinant isolates in RNA3 are infrequent (detected at a 7% rate in 3 of the 11 sampled populations) and do not become established in CMV populations. These results clearly parallel our own observations on the high and low frequency of recombinant isolates detected in the PVY and CMV datasets, respectively. Another point of interest is that among the PVY isolates detected as recombinant are the NTN (tuber necrotic) isolates, which represent a new pathotype of PVY. All these isolates were found to be tagged by the same recombination event in the CP gene, an indication that the NTN character is likely to have been acquired either after or as a result of the recombination. Similarly, the results of Kerlan *et al.* (1997) indicate that the PVY isolates belonging to new, minor groups of the virus, discriminated by unusual combinations of serological and pathogenicity properties, such as the W and Z groups, are likely to derive from recombination events. These results point to the possibility of creation of new viral pathotypes as a result of recombination.

Although it is difficult to speculate as to the reasons behind the widely different recombinant isolate frequencies found here, some hypotheses can be proposed. In the case of CMV, there seems to be strong counter-selection against inter-subgroup recombination (Fraile *et al.*, 1997), the events that would have been the clearest to detect. In the case of potyviruses, a possibility is that the distribution of isolates may have prevented recombination through geographical isolation. Such is for example the case for PPV, a virus for which the two major serotypes have apparently been geographically isolated until very recently. In fact, PPV ò6, the only known recombinant PPV isolate (Cervera *et al.*, 1993) was isolated from Yugoslavia, one of the few regions where both serotypes are known to have cohabited for a period of time.

The present results should therefore be taken as an indication that if some viruses may have already reached their equilibrium in terms of exploring their intragenetic recombination potential, this may not be the case for others, with potential consequences on the long term impact of transgenic plants expressing viral sequences. In this respect, introduction in a region from which they are absent of viral sequences under the guise of a transgene is likely to carry risks in a fashion similar to the risks caused by the introduction of a non-indigenous viral isolate. Such a situation should therefore be avoided as much as possible by using local isolates to produce transgenic crops and by the use of transgenes minimizing recombinations risks (Greene and Allison, 1996).

References

Bussière F, Lafontaine D, Perreault J-P (1996) Compilation and analysis of viroid and viroid-like RNA sequences. Nucl Acids Res 24:1793–1798

Cervera MT, Riechmann JL, Martin MT, Garcia JA (1993) 3'-terminal sequence of the plum pox virus PS and ò6 isolates : evidence for RNA recombination within the potyvirus group. J Gen Virol 74:329–334

Chenault KD, Melchers U (1994) Phylogenetic relationships reveal recombination among isolates of cauliflower mosaic virus. J Mol Evol 39:496–505

Clark DK, Duarte EA, Moya A, Elena SF, Domingo E, Holland J (1993) Genetic bottlenecks and population passages cause profound fitness differences in RNA viruses. J Virol 67:222–228

Falk BW, Bruening G (1994) Will transgenic crops generate new viruses and new diseases? Science 263:1395–1396

Felsenstein J (1989) PHYLIP – Phylogeny Inference Package (version 3.2). Cladistics 5:164–166

Fraile A, Alonso-Prados JL, Aranda MA, Bernal JJ, Malpica JM, Garcia-Arenal F (1997) Genetic exchange by recombination or reassortment is infrequent in natural populations of a tripartite RNA plant virus. J Virol 71:934–940

Gal S, Pisan B, Hohn T, Grimsley N, Hohn B (1992) Agroinfection of transgenic plants leads to viable cauliflower mosaic virus by intermolecular recombination. Virology 187:525–533

Greene AE & Allison RF (1994) Recombination between viral RNA and transgenic plant transcripts. Science 263:1423–1425

Greene AE, Allison RF (1996) Deletions in the 3' untranslated region of cowpea chlorotic mottle virus transgene reduce recovery of recombinant viruses in transgenic plants. Virology 225:231–234

Hammond R, Smith DR, Diener TO (1989) Nucleotide sequence and proposed secondary structure of Columnea latent viroid: a natural mosaic of viroid sequences. Nucl Acids Res 17:10083–10094

Higgins DG, Bleasby AJ, Fuchs R (1992) CLUSTAL V : improved software for multiple sequence alignment. CABIOS 8:189–191

Hsu Y-H, Chen W, Owens RA (1994) Nucleotide sequence of a hop stunt viroid variant isolated from citrus growing in Taiwan. Virus Genes 9:193–195

Keese P, Symons RH (1985) Domains in viroids: evidence of intermolecular RNA rearrangements and their contribution to viroid evolution. Proc Natl Acad Sci USA 82:4582–4586

Kerlan C, Tribodet M, Glais L, Astier-Manifacier S, Robaglia C (1997) Polymorphisme de la région 5' de l'ARN viral et critères de différenciation des souches du virus Y de la pomme de terre. VIèmes Rencontres de Virologie Végétale CNRS-INRA, p5

Maynard-Smith J (1992) Analyzing the mosaic structure of genes. J Mol Evol 34:126–129

Revers F, Le Gall O, Candresse T, Le Romancer M, Dunez J (1996) Frequent occurrence of recombinant potyvirus isolates. J Gen Virol 77:1953–1965

Sawyer S (1989) Statistical tests for detecting gene conversion. Mol Biol Evol 6:526–538

Tepfer M (1993) Viral genes and transgenic plants. What are the potential environmental risks? Bio/Technology 11:1125–1131

Different Mechanisms of Homologous and Nonhomologous Recombination in Brome Mosaic Virus, a Model Plant RNA Virus

Jozef J. Bujarski
Plant Molecular Biology Center and the Department of Biological Sciences, Northern Illinois University, DeKalb, Illinois 60115, USA

Abstract
Brome mosaic bromovirus, a tripartite, positive-stranded RNA virus of plants, can generate both homologous and nonhomologous intersegment RNA recombinants *in vivo*. The use of specially designed BMV RNA recombination vectors has demonstrated that recombination signals in the RNAs were different for these two recombination types. Specifically, nonhomologous recombination requires the formation of local double-stranded regions between the recombining RNAs while homologous recombination is facilitated by AU-rich sequences in conjunction with upstream GC-rich regions, common in the RNA substrates. These features most likely reflect different mechanisms involved in homologous and nonhomologous crossovers for this RNA virus.

To study if BMV replicase is involved in RNA recombination, viable mutants in both 1a and 2a replicase polypeptides were tested. Specifically, mutations within the helicase domain of 1a affected the location of crossover sites in nonhomologous recombination. This demonstrates that 1a protein is involved in nonhomologous recombination, most likely in unwinding double-stranded regions between the recombining RNAs. Mutations in the core domain of 2a, the RNA polymerase component, have reduced nonhomologous recombination below the level of detection, without affecting noticeably homologous recombination. All these observations confirm the existence of differences in the molecular mechanisms of both recombination types.

Discovery of RNA recombination in brome mosaic virus

RNA-RNA recombination was first observed for picornaviruses (Hirst, 1962; Ledinko, 1963; Cooper, 1977) and later in coronaviruses and in bromoviruses (Lai *et al.,* 1985; Ahlquist, 1992). The type member of the bromoviruses, brome mosaic virus (BMV), turned out to be highly suitable for recombination studies, because of its small size, the tricomponent nature of its positive-stranded genome, and the available full-length cDNA clones (Ahlquist, 1992). It was observed that a viable short deletion within the conserved 3'-terminal tRNA-like structure in the BMV RNA3 component was repaired *in planta* due to intraviral RNA-RNA recombination events with either of the nonmutated genomic RNA1 or RNA2 components (Bujarski and Kaesberg, 1986). Subsequent studies demonstrated that RNA recombination occurs rather frequently in bromoviruses in several different host systems, such as barley protoplasts (Rao and Hall, 1990; Ishikawa *et al.*, 1991; Nagy and Bujarski, 1992), *Chenopodium hybridum* and *C. quinoa* (Nagy and Bujarski, 1992; Rao *et al.*, 1990) and whole barley or tobacco plants (Allison *et al.*, 1990; Bujarski and Dzianott, 1991; Nagy and Bujarski, 1992). These studies revealed that both homologous and nonhomologous types of intraviral recombinants can be generated (Bujarski *et al.*, 1994; Bujarski and Nagy, 1994; Simon and Bujarski, 1994).

RNA determinants of nonhomologous recombination

During the initial experiments we observed the lack of major recombination hotspots, a relatively low recombination frequency, and the existence of strong selection pressure favoring viable 3' recombinants (Rao *et al.*, 1990; Bujarski and Dzianott, 1991; Nagy and Bujarski, 1992). These difficulties were overcome later by constructing a non-competitive RNA3-based recombination vector (designated PN0 RNA3), which was stable (i.e. did not recombine) in plants when co-inoculated with wt RNA1 and RNA2 (Nagy and Bujarski, 1993). Insertion of recombinationally active sequences into the PN0 vector, however, induced recombination with high frequency, thus allowing for targeting the junction sites in both RNAs. This vector system was utilized to study signal sequences of recombination. Comparison of crossover sites for several nonhomologous recombinants suggested the involvement of short base-paired regions between the RNA substrates in selection of crossover sites (Bujarski and Dzianott, 1991; Nagy and Bujarski, 1992). An RNA1-derived 66 nt sequence was inserted into the cloning site of PN0 RNA3 in complementary orientation (Nagy and Bujarski, 1993). This induced RNA1/RNA3 recombinants with high frequency. On the contrary, insertion of the corresponding region in direct orientation did not give recombinants at detectable frequency. The junction sites clustered (were targeted) within the heteroduplexed region (Nagy and Bujarski, 1993). These studies supported crucial evidence for heteroduplex-based nonhomologous recombination.

The above heteroduplex-driven recombination system has allowed us to study RNA sequence/structure and protein (replicase) requirements of nonhomologous recombination. Namely, a series of PN0 derivatives with RNA1-derived complementary inserts revealed that longer than 30 nt heteroduplexed regions can support recombination at high frequency. Also, analysis of junction sites revealed that they were clustered at one side of the heteroduplex region. A model which explains such characteristics proposes that the replicase enzyme (RdRp) has difficulties to pass through the heteroduplex region. Indeed, destabilizing the left-side of the heteroduplex region shifted the junction sites into the energetically more stable parts of the duplex (Nagy and Bujarski, 1993). The majority of recombinants had the junctions juxtaposed at or near counterpart nucleotides within the heteroduplex, while others had the donor and the acceptor sites farther apart from each other. This can be explained by temporary formation of nonstandard or alternative double-stranded structures just before or during the template switching (TS) events.

According to our model (Nagy and Bujarski, 1993), the formation of heteroduplexes between the RNA templates can facilitate TS events due to (i) bringing the RNA substrates into close proximity within the cells, and/or (ii) the structure or the stability of the heteroduplex regions which cause a road block, slowing down or stalling the approaching replicase at or within the double-stranded region. It has been proposed that the replicase switches between templates in a processive manner by sliding under the heteroduplex region towards the opposite side of the acceptor RNA (Nagy and Bujarski, 1993; Bujarski and Nagy, 1994). In essence, the outcome is analogous to looping out events at strong secondary structures during RNA replication.

RNA determinants of homologous recombination

Our analyses revealed several important differences between homologous and nonhomologous recombinants, including the selection of crossover sites (regions), the precision of the crossover events and the frequency of their generation. Homologous recombinants were isolated 5–10 times more frequently than nonhomologous ones (Nagy and Bujarski, 1992; Bujarski and Nagy, 1994). We concluded that in BMV the sequence and/or structure requirements of homologous recombination are probably different from those of nonhomologous recombination.

An RNA3-based recombination vector (designated PN100), similar to the above described PN0, was constructed (Nagy and Bujarski, 1993). Thereafter, a 60 nt RNA2-derived region (designated region R) was inserted into PN100 RNA3 in direct orientation. The resulting PN-H26 RNA3 construct recombined efficiently with wt RNA2, with the majority of crossovers occurring precisely within the common regions (Nagy and Bujarski, 1995). A 15 nt sequence derived from region R supported efficient RNA2/RNA3 recombinants, while 5 and 9 nt long common regions supported much more reduced levels of recombination (Nagy and Bujarski, 1995).

Mismatch mutations were introduced into region R in RNA3 to provide markers for localization of junction sites. Surprisingly, the predicted secondary structures in region R in the above mismatch-mutants, the observed homologous recombination frequencies and the distribution of junction sites (recombination profile) did not correlate, arguing that the secondary structures at the hotspot regions play limited, if any, role in homologous recombination (Nagy and Bujarski, 1995). This represented another piece of evidence that homologous and nonhomologus recombination utilize different mechanisms.

Imprecise homologous recombinants having mismatched, deleted, inserted or nontemplated nucleotides within the crossover region were also isolated (Nagy and Bujarski, 1995). These changes were almost invariably located in the crossover region, suggesting strongly that they were introduced during the crossover events, and that they point at the actual crossover sites (Nagy and Bujarski, 1995). The imprecise crossovers were concentrated within two short stretches that contained mainly A and U nucleotides (Nagy and Bujarski, 1995). These observations led to a working model which predicted that the AU-rich stretches can cause either misalignment between the RNA templates, replicase stuttering (nontemplated addition of several nucleotides), nucleotide misincorporation or their different combinations (Nagy and Bujarski, 1995 ; Nagy and Bujarski, 1996). This model was tested by increasing the length of one of the AU-rich stretches in both RNA2 and RNA3, which resulted in a remarkable increase in isolation of imprecise recombinants (as high as 86% of all homologous recombinants can be imprecise) (Nagy and Bujarski, 1996).

To study the role of AU-rich sequences during the homologous crossover events, approximately 30 nt artificial AU-rich sequences were introduced 3' to region R', which corresponded to the 5' 23 nt of region R, into both RNA2 and RNA3 vectors. This increased the frequency of homologous recombination significantly. In contrast, inserting one of the AU-rich sequences 5' to region R' neither increased the frequency of crossovers significantly nor changed the distribution of crossovers (Nagy and Bujarski, 1997). Surprisingly, however, the AU-rich sequences alone in both RNA2 and RNA3 supported only low a level of homologous recombination (Nagy and Bujarski, 1997), indicating that the 5' R' and the 3' AU-rich sequence must be present on both RNAs for efficient homologous recombination to occur.

Replacing R' sequences in both RNAs with different, both artificial and virally-derived sequences (Nagy and Bujarski, 1997) demonstrated that several unrelated GC-rich sequences increased the frequency of homologous recombination, when compared to AU-rich sequences alone. We postulate that these sequences act as recombination activators: the downstream AU-rich region is the "core" region, and the upstream GC-rich region is the "enhancer" region. According to a model proposed to explain these data, the 5' GC-rich (or average) and the 3' AU-rich sequences play different roles in the donor and acceptor RNAs (Nagy and Bujarski, 1997). The AU-rich sequences in the donor RNA may cause replicase pausing and/or the release of the enzyme, while the GC-rich upstream portion may facilitate the interaction on the complementary region of the acceptor RNA. The AU-rich sequence in the acceptor RNA may facilitate the opening of the partially double-stranded replication intermediate (RI) via local bubble structure formation.

Further evidence supporting the above model was obtained by testing sequences flanking the above recombination activators. This revealed that long (between 20-30 nt) GC-rich sequences located only at downstream positions (but not in upstream positions) in both RNA2 and RNA3 or only in RNA2, reduced significantly the incidence of recombination at upstream locations (Nagy and Bujarski, unpublished). Thus, basically the downstream GC-rich sequences acted as homologous recombination silencers on the upstream hotspot regions. Based on our model, the downstream GC-rich sequence can inhibit the formation of the bubble structure in the acceptor RNA. Also, re-initiating the strand synthesis on the acceptor RNA may be more difficult due to the close downstream location of the GC-rich region (Nagy and Bujarski, unpublished).

Effect of mutations in replicase proteins on homologous and nonhomologous recombination

The above studies gave us highly efficient recombination systems with which the role of replicase proteins could be tested. BMV genomic RNAs with mutations within the corresponding open reading frames of 1a and 2a replicase proteins were utilized. Three viable 1a protein mutants with two amino-acid insertions within the putative helicase region were tested and showed different alterations in the distribution of recombination sites and/or variable effects on the incidence of nontemplated nucleotides (mainly U residues) at recombination sites (Nagy *et al.*, 1995). One of the mutants, which was temperature sensitive in replication (designated PK19 RNA1), showed a characteristic 5' shift in crossover sites towards energetically less stable portions of the RNA1-RNA3 heteroduplex at the elevated (but still permissible) temperature (Nagy *et al.*, 1995). These results suggested that the helicase domain of 1a participates in heteroduplex-mediated crossover events.

A single amino acid mutation within the core domain of the polymerase, protein 2a, (designated DR7) inhibited the frequency of nonhomologous recombination below the level of detection (Figlerowicz *et al.*, 1997). In contrast, the frequency of homologous events was comparable in infections with DR7 or wt RNA2. Also, two viable mutations (designated MF-II and MF-V) in the N-terminal portion of the 2a protein, which is known to interact with the C-terminal domain of 1a (Kao and Ahlquist, 1992), were tested in recombination. Like for DR7, the heteroduplex-mediated nonhomologous recombinants were generated at reduced frequency in infections with either MF-II or MF-V, as compared to wt RNA2. For homologous recombination, MF-II generated recombinants at a high frequency, but MF-V reduced markedly the homologous recombination frequency (Figlerowicz, Nagy, Tang, Kao and Bujarski, unpublished). Thus, the nature and characteristics of both homologous and nonhomologous recombinants were influenced by mutations in the polymerase gene of BMV. This gives crucial support for a replicase-driven template-switching (TS) mechanism operating in both systems.

It was more difficult to find RNA2 mutants that supported reduced frequency of homologous recombination than those inhibiting the frequency of nonhomologous recombination, when compared to wt RNA2. We proposed the term "recombinosome" to describe a complex between the recombining RNAs, the replicase and other (putative) factors involved in recombination (Figlerowicz *et al.*, 1997). The stability of the recombinosome (the time required for TS events before the disassembly of the recombinosome takes place) is likely to be influenced by both the RNA components and the replicase in a different way for both recombination types.

In conclusion, our data show that different domains of the polymerase 2a protein are responsible for and/or affecting differently BMV replication, nonhomologous recombination and homologous recombination. This opens up the possibility to engineer recombinant RNA viruses which will recombine at reduced (or non-detectable) level, thus increasing the stability as well as safety of viral expression vectors.

Future experiments will be aimed at describing the function and structure of putative recombinosomes in both the homologous and nonhomologous systems in BMV. Also, the molecular components of the recombinosomes will be defined biochemically. Similar studies with other RNA viruses are needed to elucidate the general or unique nature of BMV recombination.

Acknowledgments

This research was supported by grants from NIH (ROI AI26769 and 3ROI AI26769), NSF (MCB-9305389 and MCB-9630794), and by the Plant Molecular Biology Center at Northern Illinois University.

References

Ahlquist P (1992) Bromovirus RNA replication and transcription. Curr Opin Genet Dev 2:71–76

Allison RF, Thompson G, Ahlquist P (1990) Regeneration of a functional RNA virus genome by recombination between deletion mutants and requirement for cowpea chlorotic mottle virus 3a and coat genes for systemic infection. Proc Nat Acad Sci USA 87:1820–1824

Bujarski JJ, Dzianott AM (1991) Generation and analysis of nonhomologous RNA–RNA recombinants in brome mosaic virus: Sequence complementarities at crossover sites. J Virol 65:4153–4159

Bujarski JJ, Kaesberg P (1986) Genetic recombination between RNA components of a multipartite plant virus. Nature 321:528–531

Bujarski JJ, Nagy PD (1994) Genetic RNA–RNA recombination in positive–stranded RNA viruses of plants. In: Paszkowski J (ed) Homologous recombination in plants. p. 1–24. Kluwer Academic Publisher, Dordrecht, The Netherlands

Bujarski JJ, Nagy PD, Flasinski S (1994) Molecular studies of genetic RNA–RNA recombination in brome mosaic virus. Adv Virus Res 43:275–302

Chamberlin MJ (1995) New models for the mechanism of transcription elongation and its regulation. Harvey Lect 88:1–21

Cooper P. (1977) Genetics of picornaviruses. In: Fraenkel–Conradt H, Wagner R (ed) Comprehensive Virology Vol. 9, pp. 133–207. Plenum Press, New York

Figlerowicz M, Nagy PD, Bujarski JJ (1997) A mutation in the putative RNA polymerase gene inhibits nonhomologous, but not homologous, genetic recombination in an RNA virus Proc Natl Acad Sci USA 94:2073–2078

Hirst GK (1962) Genetic recombination with Newcastle disease virus, polioviruses, and influenza virus. Cold. Spring Harbor Quant. Biol. 27:303–309

Ishikawa M, Kroner P, Ahlquist P, Meshi T (1991) Biological activities of hybrid RNAs generated by 3'–end exchanges between tobacco mosaic and brome mosaic viruses. J Virol 65:3451–3459

Kao CC, Ahlquist P (1992) Identification of the domains required for direct interaction of the helicase–like and polymerase–like RNA replication proteins of brome mosaic virus. J Virol 66:7293–7302

Lai MMC, Baric RS, Makino S, Keck JG, Egbert J, Leibowitz JL, Stohlman SA (1985) Recombination between nonsegmented RNA genomes of murine coronaviruses. J Virol 56:449–456

Ledinko N (1963) Genetic recombination with poliovirus type 1: studies of crosses between a normal horse serum–resistant mutants of the same strain. Virology 20:107–119

Nagy PD, Bujarski JJ (1992) Genetic recombination in brome mosaic virus: effect of sequence and replication of RNA on accumulation of recombinants. J Virol 66:6824–6828

Nagy PD, Bujarski JJ (1993) Targeting the site of RNA–RNA recombination in brome mosaic virus with antisense sequences. Proc Natl Acad Sci USA 90:6390–6394

Nagy PD, Bujarski JJ (1995) Efficient system of homologous RNA recombination in brome mosaic virus: Sequence and structure requirements and accuracy of crossovers. J Virol 69:131–140

Nagy PD, Bujarski JJ (1996) Homologous RNA recombination in brome mosaic virus: AU–rich sequences decrease the accuracy of crossovers. J Virol 70:415–426

Nagy PD, Bujarski JJ (1997) Engineering of homologous recombination hotspots with AU–rich sequences in brome mosaic virus. J. Virol 71: 1294–1306

Nagy PD, Dzianott A, Ahlquist P, Bujarski JJ (1995) Mutations in the helicase–like domain of protein 1a alter the sites of RNA–RNA recombination in brome mosaic virus. J Virol 69:2547–2556

Rao ALN, Hall TC (1990) Requirement for a viral trans–acting factor encoded by brome mosaic virus RNA–2 provides strong selection *in vivo* for functional recombinants. J Virol 64:2437–2441

Rao ALN, Sullivan BP, Hall TC (1990) Use of *Chenopodium hybridum* facilitates isolation of brome mosaic virus RNA recombinants. J Gen Virol 71:1403–1407

Simon AE, Bujarski JJ (1994) RNA–RNA recombination and evolution in virus infected plants. Annu Rev Phythopathol 32:337–362

Studies on RNA Recombination *in vivo* and *in vitro*

Anne E. Simon, Peter D. Nagy and Clifford D. Carpenter

Department of Biochemistry and Molecular Biology, University of Massachusetts, Amherst MA 01003, USA

Abstract

RNA recombination in the TCV system occurs between satellite (sat–) RNAs and between sat–RNAs and the TCV genomic RNA. Plants and protoplasts inoculated with TCV genomic RNA and its associated sat–RNA D accumulate recombinant molecules containing sat–RNA D at the 5' end and TCV sequence at the 3' end. We have previously identified a hot–spot for RNA recombination at the base of a stable hairpin in the 3' untranslated region of the TCV genomic RNA. Sequences comprising this hairpin and other recombination hot–spot hairpins resemble sequences located at the 5' ends of TCV and its associated subviral RNAs. Our previous model suggests that recombination occurs by template switching of the TCV replicase during plus–strand synthesis and that reinitiation of synthesis on the acceptor TCV genomic RNA occurs at sequences and/or structures that resemble sequence/structural requirements for normal plus–strand synthesis.

Others (Miller *et al.*) have speculated that the origin of some viruses may have occurred by replicase template–switching at subgenomic RNA promoters, which would also likely occur during plus–strand synthesis. Within a 600 base region that contains both the 1.45 and the 1.7 subgenomic promoters, we have found that 56% of crossover sites are clustered within 5 bases of either of the two transcription start sites, supporting the hypothesis that recombination at subgenomic promoters may have resulted in the formation of new virus species.

To further investigate requirements for recombination between sat–RNA C and sat–RNA D, we have developed an *in vitro* RNA recombination system. The crossover site in sat–RNA C is at the base of a stable hairpin that is absolutely required for recombination. We have studied the need for the sat–RNA C hairpin, the importance of a duplex region between sat–RNA D and sat–RNA C, and the need for a spacer region between the duplex and the hairpin. Our results to date indicate that (1) mutations in the hairpin that eliminate recombination *in vivo* also markedly decrease the synthesis of recombinant products *in vitro*; (2) the 3' terminal region of the duplex plays a role in terminal extension; (3) mismatches in the duplex can affect both the amount of product and the site of terminal extension; (4) the length of the spacer between the hairpin and the duplex influences the efficiency of terminal elongation. These *in vitro* data provide direct evidence for a replicase–mediated template switching mechanism for RNA recombination in TCV.

Introduction

One of the major advancements in the protection of crops against single-stranded RNA viruses was the discovery that constitutive expression of portions of a virus's genome can protect the plant against homologous, and occasionally, related viruses. Such pathogen-derived resistance is applicable to a wide range of viruses and likely involves several different mechanisms such as gene silencing or inhibition of virus uncoating (Baulcombe, 1996). Since some mechanisms of resistance require the expression of high levels of viral RNA fragments, concerns have been raised that RNA recombination between such fragments and a still pathogenic invading virus could lead to the evolution of new virus strains. In substantiation of such concerns, Schoelz and Wintermantel (1993) and Greene and Allison (1994) demonstrated that recombinants generated between transgene transcripts and an inoculated viral genome could be detected given strong selective pressure. However, others (Falk and Bruening, 1994) raised the point that plants are frequently infected by multiple viruses and suggested that cellular conditions that might lead to rapid virus evolution by RNA recombination is a natural occurrence. A number of plant and animal viruses display evidence of past recombination with other virus genomes and, less frequently, with host RNAs (Lai, 1992; Simon and Bujarski, 1994; Buck, 1996; Meyers and Thiel, 1996). However the frequency of such recombination events is unknown given that the vast majority of recombinant viruses are not viable.

In order to better estimate the risks of recombination leading to new virus species, it is important to understand why replication of viral RNAs subjects them to recombination. Recombination is widely thought to be a consequence of a loss of processivity by the viral-encoded RNA-dependent RNA polymerase (RdRp), halting complementary strand synthesis of one parental RNA and, together with the nascent strand, reinitiating synthesis on a second parental template (for a recent review see Nagy and Simon, in press). For non-homologous recombination, where the parental RNAs are not substantially related in the crossover region, analysis of sequences at junction sites reveal a diversity of mechanisms for crossover site selection. For example, non-homologous recombination in brome mosaic virus (BMV) occurs upstream of regions of that can form heteroduplexes between parental templates (Nagy and Bujarski, 1993), while recombination in flockhouse virus occurs within regions of putative secondary structure (Li and Ball, 1993). Crossover sites in tombusvirus recombinants occur primarily at the 5' termini of donor templates and single stranded regions at the base of hairpins and can also be strongly influenced by sequence homology and sequence complementarity between the parental RNAs (White and Morris, 1995). Hot-spot sequences presumptively contained within similar secondary structural elements may be sites for polymerase pausing leading to crossing over between Sindbis virus RNA templates (Hajjou *et al.*, 1996). Although there is considerable indirect evidence for the involvement of the viral RdRp in RNA recombination, recent *in vitro* studies using Qb bacteriophage RNAs suggest that a splicing-like transesterification reaction mediated by the RNAs may play a role in some non-homologous RNA recombination (Chetverin *et al.*, 1997).

The turnip crinkle virus system

We have developed whole plant (turnip and *Arabidopsis*), protoplast, and *in vitro* systems for studying sequences and structures involved in replication and recombination of RNAs associated with turnip crinkle virus (TCV). TCV (4054 bases) is among the smallest and simplest of the single component RNA viruses and is associated with numerous small, dispensable subviral RNAs (194 to 356 bases). Since subviral RNAs use the virus-encoded RNA-dependent RNA polymerase (RdRp) for replication, it is possible to use these non-protein-coding RNAs as templates in replication and recombination studies. RNA recombination occurs between satellite (sat-) RNAs and between sat-RNAs and the TCV genomic RNA.

Plants and protoplasts inoculated with TCV genomic RNA and sat-RNA D (194 bases) accumulate recombinant molecules containing sat-RNA D at the 5' end and TCV sequence at the 3' end (Nagy and Simon, 1997). These recombinant molecules are only biologically active in very rare cases (Simon and Howell, 1986; Carpenter *et al.*, 1995; Carpenter and Simon, 1996), and therefore populations of recombinants are not influenced by post-recombination selection and contain the initial crossover sites. We have identified a hot-spot for RNA recombination at the base of a stable hairpin located just downstream from the coat protein open reading frame (ORF) in the 3' untranslated region of the TCV genomic RNA (Carpenter *et al.*, 1995). Sequences comprising this hairpin (and other recombination hot-spot hairpins) resemble sequences located at the 5' ends of TCV viral and subgenomic (sg) RNAs, as well as TCV-associated subviral RNAs (Simon and Bujarski, 1994). Our working model suggests that recombination occurs when the TCV replicase switches templates during synthesis of sat-RNA D plus-strands, and that reinitiation of synthesis on the acceptor TCV genomic RNA occurs at sequences and/or structures that resemble sequence/structural requirements for normal plus-strand synthesis.

Recombination at transcript start sites for subgenomic RNA synthesis

Others (Miller *et al.*, 1995) have speculated that the generation of some viruses occurred by replicase template-switching at sgRNA promoters, which are located internally on minus-strands. This hypothesis is based on the modular nature of some viruses, which appear to have ORFs originating from several different virus groups. Although recombination between sat-RNA D and TCV primarily takes place just downstream from the ORF of the coat protein, approximately 28% of crossover junctions occur upstream of this location at two weeks postinoculation of turnip plants with the percentage of upstream recombinants decreasing at later times in the infection (Carpenter and Simon, 1996a). To determine if some of these upstream crossover sites occur in the vicinity of the TCV 1.45 kb and 1.7 kb sgRNA transcription start sites, we analyzed crossover junctions in a 550 base region extending from TCV positions 2181 through 2744. This region encompasses the promoters and transcription start sites for both sgRNAs (transcription start sites: 1.7 kb sgRNA, position 2333; 1.45 kb sgRNAs, position 2607) (Wang and Simon, in press). Using primers specific to sat-RNA D and downstream from the 1.45 kb sgRNA promoter in TCV, a number of recombinants generated in protoplasts were amplified by PCR and contained crossover sites within the specified region. Fifty-six percent of these recombinants had crossover sites clustered within 5 bases of the two transcription start sites (C.D. Carpenter and A. E. Simon, unpublished). This finding supports the hypothesis that recombination at sgRNA promoters may have resulted in the formation of new virus species and strengthens our model suggesting that recombination occurs during plus-strand synthesis in the TCV system.

Studying sequences and structures required for recombination *in vitro*

Recombination between sat-RNA D and sat-RNA C has also been studied. Recombinant molecules contain sat-RNA D at the 5' end and sat-RNA C at the 3' end. All crossover sites in sat-RNA C are at the base of a stable 30-base hairpin (designated motif1-hairpin) that is absolutely required for recombination *in vivo*. An 8-base sequence within the motif1-hairpin is identical to a portion of a non-contiguous 14-base sequence located in the 5' terminal 100 bases of sat-RNA C minus-strands, which was recently identified as one of two independent promoters for synthesis of plus-strands (H. Guan and A. E. Simon, in preparation). Such substantial sequence similarity is unlikely to be random, and may denote a recognition site for the TCV RdRp.

To further investigate requirements for recombination at the motif1-hairpin in sat-RNA C, we developed an *in vitro* system that mimics RNA recombination between sat-RNA D and sat-RNA C. This system is based on our model that suggests that the TCV replicase, after synthesizing full-length or near full-length nascent sat-RNA D plus-strands, switches to sat-RNA C minus-strand template at the base of the motif1-hairpin and then transcribes sat-RNA C plus-strand as an extension to the previously synthesized sat-RNA D plus-strand. To reconstruct recombination *in vitro*, a chimeric RNA was generated containing a 5' portion that served as a template and a 3' portion that served as a primer in a self-priming, 3'-end extension reaction (P.D. Nagy and A.E. Simon, unpublished). The 5' region consisted of minus-strand sat-RNA C sequence including the motif1-hairpin and sequences flanking both sides of the hairpin. The natural recombination hot-spot is at the 3' base of the motif1-hairpin. The 3' region of the chimeric RNA contained a short 3' proximal segment of sat-RNA D plus-strands with position -14 (relative to the natural 3' end) as the 3'-terminal nucleotide; this sat-RNA D position is a crossover hot-spot in sat-RNA D/sat-RNA C recombinants *in vivo*.

The chimeric construct supported 3'-terminal extension *in vitro* using a partially purified TCV RdRp preparation and the sat-RNA C-derived segment as the template. Transcription initiated at or close to the base of the motif1-hairpin. Thus, in this 3'-terminal extension reaction, both the donor and acceptor sites represent the recombination hot-spots observed *in vivo* between sat-RNA C and sat-RNA D. Deletion of all or part of the motif1-hairpin reduced 3'-terminal extension to near undetectable levels (P.D. Nagy and A.E. Simon, unpublished), demonstrating that the motif1-hairpin is required for efficient 3'-terminal extension *in vitro* as well as recombination *in vivo* (Cascone *et al.,* 1993). In addition, mutations within the motif1-hairpin that reduced recombination *in vivo* also dramatically reduced 3' end extension *in vitro* (P.D. Nagy and A.E. Simon, unpublished). Formation of a duplex region between the sat-RNA D-derived sequence and the sat-RNA C-derived sequence 3' of the motif1-hairpin greatly facilitated 3'-terminal extensions *in vitro*, and disruption of the putative duplex region *in vivo* also reduced the frequency of recombination between the sat-RNAs. Overall, these experiments demonstrate that the requirements of 3'-terminal extension with our chimeric RNA construct *in vitro* mimic the requirements for RNA recombination between sat-RNA C and sat-RNA D *in vivo*. Also, these experiments defined two separate elements as required for these activities: (1) a duplex region thought to bring the 3'-terminus close to the motif1-hairpin; and (2) the motif1-hairpin thought to recruit the RdRp.

Recombination as a mechanism for repair of 3'-terminal deletions

The ability of an RdRp to recombine RNAs may have arisen as a means of repairing virus genomes (Lai, 1992). Single-stranded 3' ends of some RNA viruses that do not have the putative protective features of a poly (A) tail or an esterified amino acid may be vulnerable to degradation by cellular exonucleases after virion uncoating, leading to deletion of important 3' end promoter sequences. The ability of RdRp to repair such damage by developing mechanisms to use similar 3' end sequences of sister RNAs, in the case of multipartite viruses, would give such viruses a competitive advantage.

A repair mechanism existing in the TCV system was discovered when plants inoculated with the genomic RNA and truncated sat-RNA D containing 3' end deletions of up to 13 bases still accumulated high levels of sat-RNA D (Carpenter and Simon, 1996b, 1996c). Sequencing the resultant sat-RNA D revealed that nearly all the "new" sat-RNA D terminated with the motif C1-2UGC1-3, which was similar or identical to the motif at the 3' end of natural sat-RNA D as well as the 3' ends of all RNAs associated with TCV (CCUGCCC-3'). In addition, new sat-RNA D derived from RNAs with deletions of at least 6 bases were further truncated to the -14 position (relative to the 3' end) and contained between 1 and 22 bases of heterogeneous "filler" sequence upstream from the terminal motif. New sat-RNA D accumulating in different plants had mostly different "filler" sequences, in some cases nontemplate nucleotides and segments from internal regions of plus-strand TCV. The new sat-RNA D molecules generated were therefore a new type of recombinant between TCV and sat-RNA D.

The selection of internal TCV sequence used as "filler" between the C1-2UGC1-3 terminal motif and the sequence upstream of the -14 position was not random (Carpenter and Simon, 1996c). Of the 20 independently generated new sat-RNA D analyzed, 11 contained one of three specific TCV sequences. The three internal TCV sequences were just upstream from a 6-base sequence that could form 5 or 6 base-pairs with the complement of the 3' terminal motif (CUGCCC-3'; complement is GACGGG-5'). Since at least one G:U base-pair could be formed, the six base downstream sequence was not identical to the 3' terminal motif, indicating that at least two steps were required to produce the new sat-RNA D (addition of the TCV segment and replacement of the terminal motif). By analyzing repair of deletions at the 3' end of sat-RNA C, which does not involve further truncations or TCV "filler" sequence, the terminal motif at the 3' end of TCV was determined to be the origin of the new motifs added to the ends of truncated sat-RNA C (Nagy *et al.*, 1997) and, most probably, sat-RNA D.

Sat-RNAs D and C require the CCUGCCC-3' terminal motif for synthesis of minus-strands (Carpenter and Simon, 1996b; Carpenter and Simon, unpublished). If the 3' end of TCV is the origin of the terminal motifs used in the repair of truncated sat-RNA D and sat-RNA C, a mechanism is required that would produce a complementary copy of the TCV motif and then use the complementary copy to initiate transcription directly on truncated sat-RNA C, or on a secondary template to create the "filler" required for repair of sat-RNA D. *In vitro* transcription assays using partially purified TCV RdRp demonstrated that a substantial amount of abortive products of 5 to 7 bases are generated using TCV genomic RNA as template (Nagy *et al.*, 1997), which we propose is the origin of the complementary copy of the terminal motif. In addition, extension of a 6-base oligoribonucleotide complementary to the terminal motif by the RdRp was equally efficient using full-length sat-RNA C or sat-RNA C containing a deletion of the 3' terminal 6 bases (Nagy *et al.*, 1997). This result indicated that base-pairing is not required between the primer and the template and supports our model for repair of sat-RNA C.

RNA recombination mediated by viral RdRp is a mechanism that promotes the generation of novel genomes that may have selective advantages over parental genomes. In addition, the advantageous use of recombination to repair damaged genomes illustrates the important role that recombination has likely played in generating the current variety of modern viruses. Whether other virus systems will be found to have sequence/structural requirements similar to those found in the TCV system is not known. However, more extensive analysis of recombination in other virus systems may reveal such elements, which, if removed from transgenically expressed viral-derived RNAs, may decrease the probability of unwanted recombination events.

Acknowledgments

Research in the lab of A.E. Simon is supported by National Science Foundation awards MCB-9315948 and MCB-9630191.

References

Baulcombe DC (1996) Mechanisms of pathogen-derived resistance to viruses in transgenic plants. Plant Cell 8:1833-1844

Buck KW (1996) Comparison of the replication of positive-stranded RNA viruses of plants and animals. Adv Virus Res 47:159–251

Carpenter CD, Simon AE (1996a) Changes in locations of crossover sites over time in de novo generated RNA recombinants. Virology 223:165–173

Carpenter CD, Simon AE (1996b) In vivo restoration of biologically active 3' ends of virus-associated RNAs by non-homologous RNA recombination and replacement of a terminal motif. J Virol 70:478–486

Carpenter CD, Simon AE (1996c) In vivo repair of deletions at the 3' end of a TCV satellite RNA in vivo may involve two abortive synthesis and priming events. Virology 226:153–160

Carpenter D, Oh J-W, Zhang C, Simon AE (1995) Involvement of a stem-loop structure in the location of junction sites in viral RNA recombination. J Mol Biol 245:608–622

Cascone PJ, Haydar TF, Simon AE (1993) Sequences and structures required for recombination between virus-associated RNAs. Science 260:801–805

Chetverin AB, Chetverina HV, Demidenko AA, Ugarov VI (1997) Nonhomologous RNA recombination in a cell-free system: Evidence for a transesterification mechanism guided by the secondary structure. Cell 88:503–513

Falk BW, Bruening G (1994) Will transgenic crops generate new viruses and new diseases? Science 263:1395–1396

Greene AE, Allison RF (1994) Recombination between viral RNA and transgenic plant transcripts. Science 263:1423–1425

Hajjou M, Hill KR, Subramaniam SV, Hu JY, Raju R (1996) Nonhomologous RNA-RNA recombination events at the 3' nontranslated region of the sindbis virus genome: hot spots and utilization of nonviral sequences. J Virol 70:5153–5164

Lai MMC (1992) RNA recombination in animal and plant viruses. Microbiol Rev 56:61–79

Li Y, Ball LA (1993) Nonhomologous RNA recombination during negative-strand synthesis of flock house virus RNA. J Virol 67:3854–3860

Meyers G, Thiel H-J (1996) Molecular characterization of pestiviruses. Adv Virus Res 47:53–118

Miller WA, Dinesh-Kumar SP, Paul CP (1995) Luteovirus gene expression. Crit Rev Plant Sci 14:179–211

Nagy PD, Bujarski JJ (1993) Targeting the site of RNA-RNA recombination in brome mosaic virus with antisense sequences. Proc Natl Acad Sci USA 90: 6390–6394

Nagy PD, Simon AE (1997) New insights into the mechanisms of RNA recombination. Virology, (in press)

Nagy PD, Carpenter CD, Simon AE (1997) A novel 3'-end repair mechanism in an RNA virus. Proc Natl Acad Sci USA 94:1113–1118

Simon AE, Bujarski JJ (1994) RNA-RNA recombination and evolution in virus infected plants. Annu Rev Phytopathol 32:337–362

Simon AE, Howell SH (1986) The virulent satellite RNA of turnip crinkle virus has a major domain homologous to the 3'-end of the helper virus genome. EMBO J 5:3423–3428

Simon AE, Nagy PD (1996) RNA recombination in turnip crinkle virus: its role in formation of chimeric RNAs, multimers, and in 3'-end repair. Sem Virol 7:373–379

Wang J, Simon AE (1997) Analysis of the two subgenomic RNA promoters for turnip crinkle virus in vivo and in vitro. Virology (in press)

White KA, Morris TJ (1995) RNA determinants of junction site selection in RNA virus determinants and defective interfering RNAs. RNA 1:1029–1040

Significance of RNA Recombination in Capsid Protein–Mediated Virus–Resistant Transgenic Plants

Richard F. Allison[1], Ann E. Greene[2] and William L. Schneider[1]

[1] Department of Botany and Plant Pathology, Plant Biology Building, Michigan State University, East Lansing, MI 48824–1312 USA

[2] USDA/ARS, Agronomy Department, 1150 Lilly Hall of Life Sciences, West Lafayette, IN 47907–1150 USA

Abstract

Concerns over the environmental release of transgenic organisms that contain genes from other species range from moral issues to the potential flow of the transgenic genes to related species through normal outcrossing. One concern unique to virus–resistant transgenic plants is that all or a portion of the viral gene may be incorporated by RNA recombination into a different plant virus that may infect the transgenic plant. RNA recombination is a fundamental viral evolutionary mechanism. It occurs when the viral replication complex switches templates during synthesis of a complementary RNA strand and results in chimeric RNA molecules that represent a combination of two or more templates. Recombination between a challenging virus and the viral transgene could result in chimeric viruses with unique phenotype.

Results from several laboratories using different viruses have demonstrated that the viral transgene is available for RNA recombination with a challenging virus. In our own work, deletions in the cowpea chlorotic mottle bromovirus (CCMV) capsid gene that limited the systemic spread of the virus were repaired through recombination with a CCMV viral transgene. Three of the seven recombinants caused symptoms on cowpeas that were distinct from wild–type symptoms.

Within this bromovirus system, we have demonstrated that transgenes that do not include the 3' untranslated region (UTR) are less likely to be involved in recombination events. The 3' UTR contains the viral replication–complex recognition site and has been included in the transgenes of numerous capsid protein mediated virus–resistant transgenic plants. Disruption or avoidance of this site in the transgene may reduce the involvement of the transgene in recombination. By random sampling viral RNAs, we have found a low percentage of chimeric RNAs that were derived from recombination events with other, presumably host, RNAs. This suggests that RNA recombination may occur randomly and while exclusion of the 3' UTR from the transgene discourages its involvement in recombination events, it does not ensure that it will not be involved.

Results and discussion

Plant viruses cause serious economic losses in today's agriculture, and in some geographical areas viruses limit the crops that can be grown profitably. Traditional control measures have included clean stock or seed programs, the control of virus vectors with pesticides, and the introduction of virus resistance genes from close relatives of the commercial cultivars. But with the concerns over the toxicity of pesticides and the limited reservoirs of resistance genes, alternative methods of controlling plant viruses are desirable.

Presently the most promising method of conferring viral resistance to susceptible cultivars is through the expression of viral genes within the susceptible plant. Pioneering work by Powell-Abel *et al.* (1986) demonstrated that expression of the tobacco mosaic tobamovirus (TMV) capsid (CP) gene provided tobacco with TMV resistance. Pathogen-derived resistance has been applied successfully to more than 30 different species with viral CP genes from numerous different viruses. This approach to providing plant virus resistance is both economically and environmentally attractive. However, this new technique provides several concerns that are specific to these CP-mediated virus-resistant transgenic plants (VRTPs).

One potential risk of VRTPs involves the possible transencapsidation of the genome of a challenging virus within the capsid protein expressed by the transgenic plant. Two scenarios are envisioned. One is that transencapsidation may assist the virus in the infection of the VRTP. While this may cause economic losses in that particular crop, no long term problems are predicted. The second scenario involves a temporary change in the insect vector of the challenging virus. In some viruses, the interaction of the virus with its vector is dependent on properties of the virion. If the transgenically derived virion specifies a vector that is different than the one characteristic of the challenging virus, then the transencapsidated genome could be introduced to a new host. In this case the virus genome is not genetically altered. Further encapsidation of the virus in its foreign host would be within its natural virion structure and it is unlikely that the host range of the virus would be permanently modified.

A second and more complex issue is that the transgenic RNA responsible for resistance to a specific virus may be involved in RNA recombination with a different challenging virus. Such events could lead to the incorporation of all or a segment of the viral transgene into the challenging virus. RNA recombination is a natural phenomenon that occurs when the viral replication complex changes templates during synthesis of a complementary RNA strand. If the new template is unique from the original, the resulting progeny RNA will be a chimeric molecule consisting of segments from the two different RNA templates. Comparisons of nucleotide sequences of plant viruses indicate that RNA recombination plays a role in RNA virus evolution. Theoretically, RNA recombination involving the transgene of a VRTP could result in the introduction of genetic variation within the challenging virus.

Experiments in our laboratory have demonstrated that the transgenic viral transcript is available at an appropriate location within the cell and in quantities adequate to support RNA recombination (Greene and Allison, 1994). These experiments utilized a mutant cowpea chlorotic mottle bromovirus (CCMV) which was incapable of systemic movement due to the deletion of approximately a third of the 3' region of the CP gene. Replication of this mutant was unaffected by this deletion. Transgenic *Nicotiana benthamiana* plants that transcribed the 3' two thirds of the CP gene plus the physically associated 3' untranslated region of CCMV were challenged with this movement-defective CCMV. The lack of an initiation codon in the viral transcript made translation unlikely, and none of the resulting transformants was resistant to CCMV. Of the 235 plants challenged, seven became systemically infected. Analysis of virus recovered from these systemically-infected plants revealed a complete CP gene containing point mutations that distinguished the transgene from the wild-type gene. Interestingly, each recombinant CP sequence was different from the others, and all were distinct from the wild-type CCMV CP gene. When passaged on the natural host, cowpeas, symptoms produced by three of the recombinants were distinct from wild-type CCMV. Similar experiments in other viral systems have demonstrated recombination involving a viral transgene (Lommel and Xiong, 1991; Gal *et al.,* 1992; Schoelz and Wintermantel, 1993). In most of these experiments, deletion mutants were restored by recombination with a transgene derived from the same virus. When assessing the risks of VRTPs, it is of little concern if recombination occurs between a viral transgene and the parental virus from which the transgene was derived. Certainly an environment suitable for recombination is supplied during each natural infection of non-transgenic plants. However, other viruses will challenge the transgenic plant in the field and recombination with theses viruses could generate chimeric viruses with unique phenotypes.

Recombination may occur with a known pathogen of the transgenic cultivar or it may involve a virus that does not systemically infect that particular cultivar. Opportunities for recombination between two known pathogens of the same plant appear to occur during mixed infections when the two viruses infect the same plant. There are numerous examples of mixed infections; while some suggest that the viruses undergo replication simultaneously within the same plant cell, other examples are not as convincing. In cases where transencapsidation between two wild type viruses occurrs, the RNA of one virus is available to the assembling virion of the other virus. This implies that the two viruses underwent replication simultaneously and therefore have ample recombination opportunities. But most reports of mixed infections record the detection of two or more viruses in a particular plant. This may represent the accumulation of viruses over the course of a growing season rather than the simultaneous infection of the plant with two different viruses. Although it is unlikely that recombination opportunities never occur during a mixed infections, the frequency of such opportunities may not be as great as first envisioned. In our attempts to establish simultaneous mixed infections between brome mosaic bromovirus (BMV) and CCMV, only thirty percent of the plants yielded mixed infections (Schneider and Allison, unpublished results). The remaining plants became infected with either BMV or CCMV. Thus a mixed infection between these two viruses was the exception rather than the norm.

There is a second point to consider when comparing recombination in natural mixed infections to recombination in VRTPs. During a virus infection, the virus replicates in the initially inoculated cell and moves from that initial point to adjacent cells, initiating and completing replication and encapsidation as it proceeds. Viral replication is not continuous within a particular cell but is transient; unecapsidated viral RNA is likely degraded as are host encoded mRNAs. Thus when a virus challenges a plant previously infected by another virus, it does not necessarily come in contact with the replication complex or the unencapsidated RNA of the previously established virus. In contrast, transcription of the viral transgene in VRTPs is controlled by a constitutive promoter which provides a continual supply of transgene transcripts in each plant cell which is available for recombination.

The second scenario involves the potential for recombination with viruses considered non–pathogens of the transgenic cultivar. Frequently these viruses are capable of replication in protoplasts derived from that same cultivar. This suggests that their non–infectious status is due to their inability to move systemically within the plant. Thus when the virus is inoculated onto a non–host, it may replicate in the initially infected cell where it remains. Consequently, even non–pathogenic viruses may have an opportunity for recombination with the viral transgene. In this case there is a high selection pressure for a recombinant virus that can overcome the movement barriers.

The CP gene, which is the most frequently used viral transgene to date, is in many cases a multifunctional gene. By definition the CP gene encodes a structural protein, but in some viruses the protein provides vector specificity, and there are reports of the CP's involvement in systemic movement (Dolja *et al.*, 1994; Hacker *et al.*, 1992; Schneider *et al.*, 1997; Suzuki *et al.*, 1991; Wellink *et al.*, 1989; Ziegler–Graff *et al.*, 1996). At this point it is impossible to predict if all or part of a capsid gene could be usefully incorporated into another virus.

As various organizations weigh the risks and benefits of genetically altered field crops, it is important to realize that the viral transgenes in today's VRTPs may be quite different from those of the future. Since these plants are relatively easy to make and the market will be quite competitive, the design of new constructs will be based on the success of the current cultivars and their transgenes. If recombination is a problem in the field, it will become readily apparent that altered versions of the transgenes are required. In the future, it may be possible to move naturally occurring resistance genes between species. In most cases this awaits the identification, isolation and characterization of such naturally occurring genes and an indication that these natural resistance genes will provide resistance in new genetic backgrounds.

References

Dolja VV, Haldeman–Cahill R, Montgomery AE, Vanderbosch KA, Carrington JC (1994) Capsid protein determinants involved in cell–to–cell and long distance movement of tobacco etch potyvirus. Virology 206:1007–1016

Gal S, Pisan B, Hohn T, Grimsley N, Hohn B (1992) Agroinfection of a transgenic plant leads to viable cauliflower mosaic virus by intermolecular recombination. Virology 187:525–533

Greene AE, Allison RF (1994) Recombination between viral RNA and transgenic plant transcripts. Science 263:1423–1425

Hacker DL, Petty ITD, Wei N, Morris TJ (1992) Turnip crinkle virus genes required for RNA replication and virus movement. Virology 186:1–8

Lommel SA, Xiong Z (1991) Reconstitution of a functional red clover necrotic mosaic virus by recombinational rescue of the cell-to-cell movement gene expressed in a transgenic plant. J Cell Biochem 15A:151

Powell-Abel P, Nelson RS, De B, Hoffman R, Rogers SG, Fraley RT, Beachy RN (1986) Delay of disease development in transgenic plants that express the tobacco mosaic virus coat protein gene. Science 232:738–743

Schneider, WL, Greene AE, Allison RF (1997) The carboxy-terminal two-thirds of the cowpea chlorotic mottle bromovirus capsid protein is incapable of virion formation yet supports systemic movement. J Virol (in press)

Schoelz JE, Wintermantel WM (1993) Expansion of viral host range through complementation and recombination in transgenic plants. Plant Cell 5:1669–1679

Suzuki M, Kuwata S, Katoka J, Masuta C, Mitta N, Takanami Y (1991) Functional analysis of deletion mutants of cucumber mosaic virus RNA 3 using an in vitro transcription system. Virology 183:106–113

Wellink J, van Kammen A (1989) Cell-to-cell transport of cowpea mosaic virus requires both the 58K/48K proteins and the capsid proteins. J Gen Virol 70:2279–2286

Ziegler-Graff V, Brault B, Mutterer JD, Simonis MT, Herrbach E, Guilley H, Richards KE, Jonard G (1996) The coat protein of beet western yellows luteovirus is essential for systemic infection but the viral gene products p29 and p19 are dispensable for systemic infection and aphid transmission. Molec Plant-Microbe Interact 9:501–510

Transgenic Plants Expressing Viral Sequences Create a Favourable Environment for Recombination Between Viral Sequences

Gábor Jakab[1], Fabián E. Vaistij[2], Eric Droz[1] and Pia Malnoë[1,2]

[1]Station Fédérale de Recherches en Production Végétale de Changins, Route de Duillier, Case postale 254, CH-1260 Nyon 1, Switzerland

[2]Université de Genève, Département de Biologie Végétale, Place de l'Université 3, 1211 Genève 4, Switzerland

Abstract

Risk assessment of transgenic plants expressing viral sequences has been mainly concerned with the direct interaction between the transgenic mRNA or the protein product and an infecting viral genome. There is another important aspect to this problem: the use of such transgenic plants may create new environments for viruses. We describe here two different situations which increase the risk of recombination between viral sequences.

The resistance which relies on homology-dependent gene silencing can be impaired in offsprings where the transgene copies have segregated. This loss of resistance is often accompanied by an increased level of the transgene product. Using such a plant, we have shown efficient complementation of a defective potato virus Y (PVY) genome by the transgenic coat protein. These data indicate that the use of virus-resistant transgenic plants synthesising a functional viral gene product might create a new environmental niche for mutated viruses. The presence of the functional protein can liberate the corresponding viral gene from the normal selective pressure and increase the opportunity for recombination between the viral genome and the transgenic viral mRNA.

Co-infection with two different strains of PVY in transgenic plants, resistant to only one of the strains, generated recombinant viruses. However, in control experiments with wild-type plants, no recombinants were detected. This shows that there is a high selective pressure for recombinants in transgenic plants where the viral resistance is based on sequence homology-dependent silencing. These results should be considered when transgenic virus-resistant plants are commercialised and introduced into agricultural systems.

Introduction

The concept of pathogen-derived resistance, when viral sequences are expressed in transgenic plants, has proved to be an extremely effective approach to obtain virus-resistant plants. The resistance of these transgenic plants can be based on at least two different mechanisms. In the first case the synthesis of a functional or a dysfunctional viral protein is necessary. Good examples include coat protein-mediated resistance to TMV (reviewed in Fitchen and Beachy, 1993) and expression of mutated viral movement proteins in plants (Cooper *et al.*, 1995). The second type of resistance is based on homology-dependent gene silencing (Baulcombe, 1996). The transgenic plants belonging to this group accumulate only low levels of transgenic products. This kind of resistance is exclusively RNA-dependent and can be obtained with untranslatable as well as translatable transgenes.

Field releases of crop plants expressing a viral gene have raised the following question: will the genetic makeup of viral populations or the epidemiology of certain viral diseases be affected by the use of these plants in our agro-ecosystem (Tepfer, 1993)? This could happen by three different processes: transcapsidation, recombination and synergism (Robinson, 1996). These three hazards all involve direct interactions between the transgene products (either RNA or protein) and the invading virus genome. Theoretical studies implied that the most serious problem would be a recombination between the transgenic mRNA and a viral genomic RNA able to infect the transgenic plant. Another event, heterologous encapsidation, is considered as less serious as the modification is not genetically transmitted (Farinelli *et al.*, 1992). The hazards arising from synergism are relatively easy to identify by experiments in containment before a release is undertaken. However, in this study, we show that we have to consider another important risk which lies in the fact that a transgenic plant creates a new environment. This environment has direct effects on virus evolution by complementation of defective viral genes or by an increased selective pressure for recombined viruses.

Virus strains used in this study

Potato virus Y (PVY), the type-member of the largest plant virus genus (potyvirus genus), has been selected for our risk assessment experiments. This virus comprises several strains classified into two main groups. The common or "O" strains occasion mottle on tobacco leaves, while the necrotic or "N" strains cause systemic veinal necrosis in tobacco. Numerous sequence data of isolates from both groups and strain-specific antibodies make PVY an excellent tool for RNA recombination experiments.

An infectious cDNA copy of the complete genome of PVY is fundamental to study RNA recombination in transgenic PVY-resistant plants. Such a cDNA has been obtained for PVY-N605 (a Swiss necrotic isolate) and put under the control of two different promoters: the CaMV 35S promoter and the SP6 promoter. With the 35S promoter, we are relying on *in vivo* transcription and translation, once the cDNA has been introduced into the plant cell either by mechanical or biolistic means. With the SP6 promoter, transcription and capping of the viral genomic RNA is carried out *in vitro*. The RNA is subsequently introduced into the cell by mechanical inoculation.

To study recombination under greenhouse conditions during a reasonable time frame, we are using what we consider to be optimal conditions for recombination. These conditions might not correspond exactly to field conditions, but they help us to predict what could happen in the fields. We have selected two strains (N and O) of PVY instead of two different virus species, because several natural examples of recombination between different strains of PVY have been reported (Revers *et al.*, 1996).

Instability of virus resistance based on homology-dependent gene silencing

Transgenic tobacco and potato plants expressing the coat protein (CP) gene and short flanking sequences of the potato virus Y (PVY) N strain have been obtained (Farinelli and Malnoë, 1993). These plants show various degrees of resistance against the different strains of PVY, and were field-tested during two years (Malnoë *et al.*, 1994). Although the CP was not detectable in healthy transgenic plants, after infection with PVY-O, synthesis of the transgenic (N) CP and heterologous encapsidation (phenotypic mixing) were observed (Farinelli *et al.*, 1992).

There is a general consensus that only transgenic plants displaying a high level of virus resistance will be released in the field on a large scale. However, this trait is not necessarily stable and the degree of silencing can be modulated by external factors. Also, hybrids of these highly resistant lines with non-transgenic cultivars of the same crop or wild relatives from neighbouring fields may result in transgenic plants with narrower resistance or sensitive plants which express the transgene at a higher level. A good example for this situation is the following experiment.

The original 4B5 tobacco plant (R0 generation) contains five copies of the transgene and shows complete resistance to both N and O strains of PVY. The region of PVY-N605 carried by this plant comprises 300 bases of the NIb gene, which precedes the CP gene in the viral genome, the CP gene and the following 200 bases from the 3'untranslated region (UTR) (Farinelli and Malnoë, 1993). After a backcross, we have obtained segregating plants (RI generation) with different degrees of resistance. Plants harbouring all five copies of the transgene (like the original 4B5) displayed high resistance to all PVY isolates tested. The transgene products, RNA and protein, were below detection levels. However, plants which had lost different copies of the transgene showed either resistance to only the homologous virus strain (e.g. RI-28 plant) or sensitivity to all PVY strains (e.g. RI-45 plant). The transgenic mRNA and, after virus infection, its protein product were only detectable in the latter case. The different characteristics of these plants are summarised in Table 1.

The resistance of 4B5 plants is based on silencing of the transgene and is therefore associated with a very low level of the transgene mRNA. A frequently used argument exists whereby the undetectable amounts of the transcript can efficiently reduce the possibility of recombination between the transgene and the viral genome. The silenced state of the transgene depends on the copy number and the location of the transgene; it can be lost by segregation and therefore cause an increase in the level of transcripts. This complicates further the ecological situations created by the release of virus-resistant plants expressing viral sequences. Hence even when the resistance is based on silencing, we have to consider the potential of the transgene insert for recombination. It seems important to avoid the use of regions that might facilitate recombination, such as sequences from sub-genomic RNA promoters or from one or the other terminus of the virus genome. If these guidelines are observed, the viral replicase must change template twice instead of once. Although this strategy is able to reduce the recombination frequency, it is not an absolute prevention.

Table 1. Resistance to PVY of tobacco line 4B5 and its progeny

Name of the clone	Resistance to:	Number of copies of the transgene	Detectable amounts of mRNA	Inducibility of the synthesis of the coat protein
4B5	all PVY strains tested (N and O)	5	-	-
RI-28	PVY-N	3	-	-
RI-45	sensitive	3	+++	+++

Complementation of a defective virus in transgenic plants

The viral SP6-cDNA used in this study contains a 37 bp deletion between two *Xho* I sites within the coat protein gene (PVY-dXho). This deletion creates a frameshift mutation and results in a dysfunctional coat protein. The transgenic recipient plant, RI-45, is sensitive to PVY. The transgene is highly expressed at the RNA level. However, the coat protein is only synthesised once the plant is infected by PVY or a related virus. The transgene does not contain the very 3' part of the UTR, which means that a double recombination event is necessary to restore the wild type CP.

RI-45 plants were infected with RNA obtained *in vitro* from the PVY-dXho cDNA construct. 19 days post inoculation (dpi) necrotic lesions as well as the presence of the N605 coat protein (by ELISA) could be detected. A northern blot with total RNA indicated that the virus had been able to replicate, and an electron microscope study showed that it had also formed complete viral particles.

In order to determine whether this virus was due to a recombination event or a heterologous encapsidation, a leaf extract was used to inoculate untransformed tobacco SR1 plants and RI-45 plants. The presence of the virus could only be detected in the RI-45 plants, which indicated that the virus was obtained by heterologous encapsidation. The fact that all three RI-45 plants became infected shows that: (1) a defective heterologously encapsidated virus can efficiently infect a plant; (2) the synthesis of the transgenic coat protein can be efficiently induced and complement the defective genomic coat protein.

Recombination between viral genomic RNA and viral transgenic mRNA

The virus has been subcultured 3 times during 4 months onto a total number of 30 RI-45 plants. The appearance of revertant viruses, obtained after recombination between the transgenic CP mRNA and the mutated viral genome, was monitored by regular transmission to tobacco SR1 plants. Despite the high titre of the defective virus genome, the first recombined virus was only observed after four months maintenance in 30 transgenic plants. These results support the earlier observation on other viruses that the 3'UTR plays a significant role in RNA recombination in transgenic plants (Greene and Allison, 1996). However, the lack of the replication initiation site in the transgene did not prevent recombination between the virus and the transgene.

In non-transgenic plants, contrarily to what happened in transgenic plants, the initial growth of the defective virus was quickly overcome by the growth of the revertant virus. Therefore, complementation by a co-infecting virus and by a transgene product have opposite consequences. Competition with the wild-type virus usually eliminates a defective genome rapidly, whilst complementation of a defective genome by a transgene supports its replication.

Liability of homology-dependent viral resistance

Discussions of the risk associated with the release of transgenic plants expressing viral sequences have been mainly concerned with the direct interaction between the transgenic mRNA or the protein product and an infecting viral genome. We have seen, however, that there is another important aspect to this problem. The use of such transgenic plants creates new environments for viruses. As we have demonstrated, the original resistant plants can give rise to sensitive plants which still express the transgene. When the resistance is based on co-suppression, the loss of resistance and the increased level of transgene product are often coincidental. In these cases, functional transgene products can efficiently complement a defective viral genome. The presence of such a product can liberate the corresponding viral gene from the selective pressure which would normally have eliminated the defective genome. These defective genomes are usually less competitive than the wild-types, therefore their complementation by viruses is unlikely, since the intact genomes can simply over-replicate the defective ones. Albeit in transgenic plants competition does not exist, therefore the defective genome can persist for a long period which increases the risk of recombination with the transgenic mRNA.

A valuable strategy to avoid complementation involves the use of non-functional inserts from the virus genome. Similarly, the frequency of recombination with the transgenic mRNA can be reduced by the introduction in the plants of viral fragments as short as possible. Following this strategy, the resistance obtained will always be based on homology-dependent gene silencing. Consequently, viruses which have eliminated the "silenced region" from their genome would be viable in these plants. Recombination seems to be the simplest way to get rid of the target sequence by replacing it with the analogous region of other strains or virus species.

In fact, in an experiment where RI-28 plants were infected simultaneously with a PVY-N and a PVY-O strain, we were able to observe recombinants in 4 out of 12 plants. This result indicates a high frequency of template switching during replication, generating a large number of viable recombinants, because of the strong sequence homology (90%) between the two strains. Sequence analysis of the recombinant virus genomes revealed that the homologous recombination between the two strains occured in two different positions. However, the sites of recombination seem predetermined: PVY strains recombined at the same sites have already been described (Revers *et al.*, 1996). No recombinant viruses were detected when SR1 plants were infected with a mixture of N and O strains of PVY. This shows the importance of the elimination of the parental virus by the transgenic plants. In this new environment the recombinant virus is more competitive than the original one.

Under usual agricultural conditions, plant viruses have many opportunities to interact genetically. Multiple infections occur commonly in both crop and weed hosts. Although recombination between different virus genomes can happen readily in the field (Simon and Bujarski, 1994), existing viruses are surprisingly stable, having evolved into fit competitors over a long time scale. Each virus, or even each strain has an optimal configuration determined by a specific environment and its selective pressure. The rate of template switching between two viral genomes in untransformed and transformed plants is probably the same, but the selective pressure is different. In transgenic virus-resistant host plants, the selective pressure favours the appearance of recombined viruses which have eliminated the "degradable target sequence". The recombined virus will be better fitted to the new situation and a new equilibrium will be reached. This way, transgenic virus-resistant plants where the resistance is based on homology-dependent silencing, favour the appearance of new recombined viruses due to a new selective pressure.

The virus-resistant cultivars developed by traditional plant breeding have given rise to new virulent virus strains (Dawson and Hilf, 1992). Generally, the virulent virus was not as competitive as the wild-type virus in the original susceptible plant. This will probably be true for the recombinant viruses developed in transgenic plants. However cultivation on a large scale of these selective transgenic plants could maintain and fix the recombinant viruses in the environment. This would result in quick loss of the resistance based on silencing.

References

Baulcombe DC (1996) Mechanisms of pathogen-derived resistance to viruses in transgenic plants. Plant Cell 8: 1833–1844

Cooper B, Lapidot M, Heick JA, Dodds JA, Beachy RN (1995) A defective movement protein of TMV in transgenic plants confers resistance to multiple viruses whereas the functional analog increases susceptibility. Virology 206:307–313

Dawson WO, Hilf ME (1992) Host-range determinants of plant viruses. Annu Rev Plant Physiol Plant Mol Biol 43:527–555

Farinelli L, Malnoë P, Collet GF (1992) Heterologous encapsidation of potato virus Y strain O (PVY^{O}) with the transgenic coat protein of PVY strain N (PVY^{N}) in Solanum tuberosum cv. Bintje. Bio/Technology 10:1020–1025

Farinelli L, Malnoë P (1993) Coat protein gene-mediated resistance to potato virus Y in tobacco: Examination of the resistance mechanisms – Is the transgenic coat protein required for protection? Mol Plant-Microbe Interact 6:284–292

Fitchen JH, Beachy RN (1993) Genetically engineered protection against viruses in transgenic plants. Annu Rev Microbiol 47:739–763

Greene AE, Allison RF (1996) Deletions in the 3' untranslated region of cowpea chlorotic mottle virus transgene reduce recovery of recombinant viruses in transgenic plants. Virology 255:231–234

Malnoë P, Farinelli L, Collet GF, Reust W (1994) Small-scale field tests with transgenic potato, cv. Bintje, to test resistance to primary and secondary infections with potato virus Y. Plant Mol Biol 25:963–975

Revers F, Le Gall O, Candresse T, Le Romancer M, Dunez J (1996) Frequent occurrence of recombinant potyvirus isolates. J Gen Virol 77:1953–1965

Robinson DJ (1996) Environmental risk assessment of releases of transgenic plants containing virus-derived inserts. Transgenic Res 5:359–362

Simon AE, Bujarski JJ (1994) RNA–RNA recombination and evolution in virus-infected plants. Annu Rev Phytopathol 32:337–362

Tepfer M (1993) Viral genes and transgenic plants. What are the potential environmental risks? Bio/Technology 11:1125–1132.

Behavior of Cucumovirus Pseudorecombinant and Recombinant Strains in Solanaceous Hosts

Mireille Jacquemond[1], Katalin Salánki[2], Isabelle Carrère[1], Ervin Balázs[2], Mark Tepfer[3]

[1] INRA, Station de Pathologie Végétale, BP 94, 84143 Montfavet Cedex, France
[2] Institute for Plant Sciences, Agricultural Biotechnology Center, POB 411, 2101 Gödöllö, Hungary
[3] INRA, Laboratoire de Biologie Cellulaire, 78026 Versailles Cedex, France

Abstract
Pseudorecombinant and chimeric recombinant strains have been created between two strains of cucumber mosaic virus (I17F–CMV and R–CMV) or between R–CMV and tomato aspermy virus (P–TAV), and their biological properties have been assessed on certain solanaceous hosts. Results involving the two CMV strains suggest that the different hosts fall into two main groups. The first corresponds to hosts for which the nature or the gravity of the symptoms is mainly determined by RNAs 1 and 2. The second group is composed of hosts for which symptomatology is mainly governed by RNA 3. In these cases, it is the 3' part of the molecule, bearing the coat protein (CP) gene, which is primarily involved. One chimeric recombinant CMV strain has distinctive properties, since it cumulates the severe symptoms induced by both parental strains in *Nicotiana glutinosa*. Pseudorecombinants and recombinants created between R–CMV and P–TAV confirmed the importance of the 3' part of RNA 3 in determining symptomatology in this host. Unexpected results were observed with one chimeric recombinant, which proved unable to infect tobacco, although it can replicate in tobacco cells. The pseudorecombinant strain composed of P–TAV RNAs 1 and 2 and RNA 3 from R–CMV induced considerably aggravated symptoms on *Nicotiana benthamiana*. This property is not related to a difference in replication efficiency.

Introduction

Both homologous recombination between similar viral RNAs and heterologous recombination between unrelated RNA sequences (from different viruses or a virus and its host) are now well documented (for a review see Simon and Bujarsky, 1994). Two types of evidence that recombination is one of the major determinants for virus evolution have accumulated during the past few years. First, sequence comparisons, and more particularly identification of consensus sequences, have strongly suggested that early recombination events have occurred between distinct viruses. For instance, the two major groups of luteoviruses are thought to have arisen by ancestral recombination events (Martin *et al.*, 1990). Second, the accumulation of sequences of numerous strains of certain viruses has made it possible to detect intra-species strain-specific recombination. For example, a systematic search for recombination within certain potyviruses has shown that many strains are the result of intra-species recombination (Revers *et al.*, 1996). In addition to true recombination, in which nucleotide sequences are covalently joined, reassortment of entire genome segments between related viruses gives rise to pseudorecombinant strains, and can thus also result in novel associations of viral genes. Within the cucumovirus group, both types of events have been observed. In the first case, a true recombination event has occurred in a pseudorecombinant created in the laboratory between cucumber mosaic virus (CMV) and tomato aspermy virus (TAV) (Fernandez-Cuartero *et al.*, 1994). In the second case, natural pseudorecombination has led to the emergence of a new cucumovirus containing RNAs 1 and 2 from peanut stunt virus (PSV) and RNA 3 from CMV (White *et al.*, 1995).

Recombination between viral RNAs naturally requires that both genomes infect simultaneously the same host, a situation that occurs frequently in the field. It is generally agreed that numerous recombination events are likely to occur when plants are co-infected by related viruses, even though essentially no recombinants are observed. In fact, this is expected since: i) only recombinants that have a selective advantage would come to dominate the population of viral RNAs and, ii) only those viruses that induce atypical or novel symptoms will be noticed, and hence analyzed. Large-scale cultivation of transgenic plants expressing viral sequences could represent another area for emergence of recombinants in nature, since the situation in transgenic plants could be similar to that in multiply infected normal plants. That recombination between a host transgene and the genome of an infecting virus can indeed occur under conditions of strong selection pressure was demonstrated as early as 1991 for RNA, then DNA viruses (Lommel and Xiong, 1991; Gal *et al.*, 1992; Schoelz and Wintermantel, 1993; Greene and Allison, 1994; Greene and Allison, 1996) and, more recently, has also been described under conditions of weaker selection pressure in favor of the recombinant (Wintermantel and Schoelz, 1996).

If recombination between viral RNAs is frequent, then it is pertinent to examine if recombinant viruses can have novel biological characterisics. Thus, one of the strategies we are developing for assessing the risks due to RNA recombination consists in creating pseudorecombinant and recombinant strains within the cucumovirus group and evaluating their biological properties. The ability of these strains constructed *in vitro* to replicate, to move and to induce symptoms in different solanaceous hosts is presented below.

Materials and Methods

Viruses and cloned cDNAs

The cucumovirus strains under study are I17F- and R-CMV, representative of subgroups I and II, respectively (Jacquemond and Lot, 1981), and P-TAV (Salánki *et al.*, 1994). Cloned cDNAs suitable for transcription *in vitro*, corresponding to the full-length RNAs 1, 2 and 3 of R-CMV and RNA 3 of I17F-CMV have been obtained as described by Hayes and Buck (1990). The sequences of the two CMV cDNAs 3 (pR3 and pI3) have been described elsewhere (Carrère, 1993). An *Eco*RV restriction site has been introduced within the intergenic region of pI3 at position 1037. Chimeric cDNAs 3, pIR3 and pRI3, were created using this site and an *Xba*I site present at position 1004 in pR3. The cDNA 3 of P-TAV (pT3) and its nucleotide sequence have been described earlier (Salanki *et al.*, 1994). For construction of chimeric CMV-TAV RNA3 molecules, an *Nde*I restriction site was introduced at the initial Met residue of the coat protein (CP) gene of pT3 and pR3 by site-directed mutagenesis, creating pR3/NdeI and pT3/NdeI. Using this *Nde*I restriction site and the *Bam*HI restriction site that is beyond the 3' end in the cDNA clones, two chimeric plasmids were created, pRT and pTR. All the cDNAs 3 are presented in Fig. 1.

Inocula and plants

The I17F-CMV strain, which contained no detectable necrogenic satellite RNA, was inoculated as viral RNA purified from tomato plants, at a concentration of 10 μg/ml. The pseudorecombinant strain containing RNAs 1 and 2 of P-TAV and RNA 3 of R-CMV (designated T1T2R3) was created by mixing extensively gel-purified viral RNAs (Salànki *et al.*, 1997). All other inocula were prepared using *in vitro* RNA transcripts. The different plasmids were digested with *Bam*HI prior to transcription. Transcription was carried out as described in Salánki *et al.* (1994). Each artificial inoculum contained approximately 2 μg of each capped RNA transcript, and was applied to young tobacco (*Nicotiana tabacum* Xanthi) plants (two or three leaf stage). Plants were kept in the greenhouse under individual cages. Viral RNA was purified from virus purified 2-3 weeks after infection. These viral RNA preparations were used as inocula, at a concentration of 10 μg/ml, for the experiments described below.

Preparation and transfection of protoplasts

Protoplasts were isolated according to Nagy and Maliga (1976) from tobacco leaves. Protoplast transfection with RNA transcripts or with purified viral RNAs was carried out by electroporation: 400 μl of a protoplast suspension of 5x105 protoplasts/ml were electroporated with 2-5 μg of each transcript or viral RNA (Saunders*et al.*, 1989).

RT-PCR-RFLP analysis of progeny CMV RNA 3

2 μg of viral RNA or 5 μg of total RNA were used in a reverse transcription assay using an oligonucleotide complementary to the 14 last nucleotides of RNA 3 of both CMV strains. Consensus oligonucleotides were designed in order to amplify two regions within cDNA 3 of each subgroup. The first corresponds to a part of the 3a movement protein (MP) gene (positions 361-385 and 629-653 for R and I17F cDNAs 3, respectively), whereas the second is located within the CP gene (positions 1505-1534 and 1787-1816 for R and I17F cDNAs 3, respectively). The PCR products were digested with enzymes that allow to distinguish the two cDNAs and analyzed on 12 or 15% polyacrylamide gels.

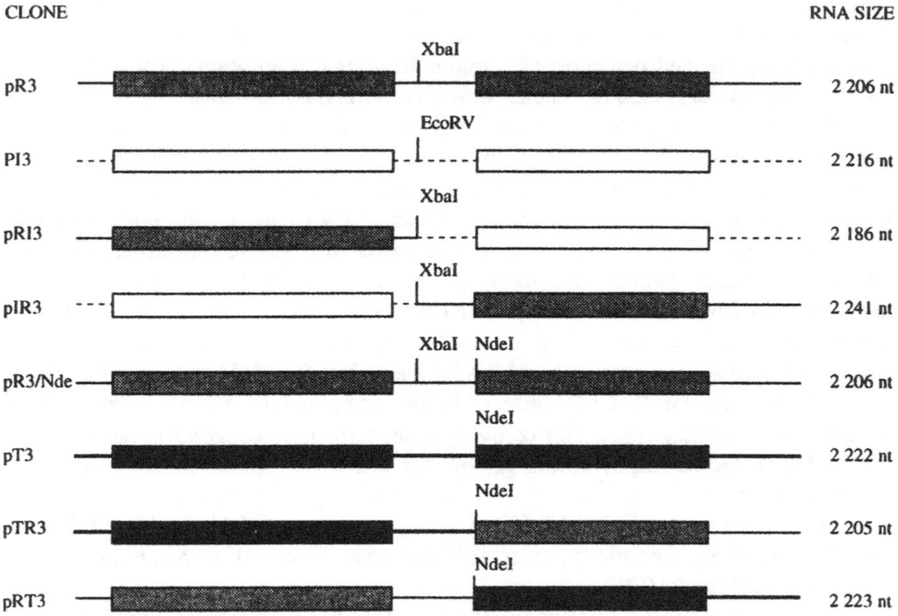

Figure 1. Schematic representation of cDNAs 3 of R and I17F-CMV and P-TAV RNAs 3, and of the different recombinants created by exchange of half parts from R-cDNA 3 and either I17F or TAV-cDNAs 3. Sequences from R-CMV are indicated by grey boxes and thin continuous lines, those from I17F-CMV by white boxes and dotted lined, those from P-TAV by black boxes and thick continuous lines. The *Xba*I site was naturally present in pR3 at position 1004. An *Eco*RV site has been created by site-directed mutagenesis in pI3 at position 1037. Filling in of each fragment generated by digestion of pR3 by *Xba*I, and ligation with the corresponding fragment produced by digestion of pI3 by *Eco*RV, created an *Xba*I site in both cases. An *Nde*I site has been created by mutagenesis at positions 1 226 and 1 225 in pR3 and pT3, respectively. In all cases, the RNA transcribed from the mutated cDNA had the same biological properties as the RNA transcribed from the native cDNA.

Analysis of plants

Total RNA was purified from infected plants or transfected protoplasts according to White and Kaper (1989). Northern analysis was carried out following electrophoresis through 1.1% formaldehyde-containing agarose gels, using random-primed ^{32}P-labelled probes, according to standard protocols (Sambrook *et al.*, 1989). The probe is specified in the text for each experiment.

Results

Biological properties of CMV pseudorecombinant and recombinant strains

Apart from I17F, all others strains were derived from artificial inocula and corresponded to RNAs 1 and 2 of the R strain plus one of the four native or recombinant CMV RNAs 3 (Fig. 1). These four strains were amplified first on young tobacco. The conformity of progeny RNA 3 was verified by RT–PCR–RFLP analysis of regions within the MP and CP genes. Although this method would have permitted detection of a minor RNA that represents 0.1% of the total RNA 3, only the expected molecules were found (data not shown).

I17F–CMV belongs to subgroup I and induced very severe symptoms on all eight hosts tested (Table 1). By contrast, R–CMV belongs to subgroup II and induced considerably weaker symptoms, except on *N. glutinosa*. Based upon the type and the gravity of the systemic symptoms they express, the eight hosts fall within two main groups. The first is composed of *Physalis floridana*, petunia, tomato, *Nicotiana glauca* and tobacco, on which I17F–CMV can be clearly distinguished from all other strains. For these hosts, RNA 3 does not bear major determinants of symptomatology. Nevertheless, minor differences are observed among the four strains composed of RNAs 1 and 2 of R–CMV and any of the four RNAs 3. In particular, R–CMV induced milder symptoms in all hosts, where only a mild mosaic was observed, with recovery from infection occurring in some cases. The order of decreasing symptom gravity for these four strains remains the same for all these hosts: I3 > RI3 > IR3 > R3. Although *N. glauca* clearly belongs to this group of hosts, since only I17F induced strong symptoms, on this host RNA 3 does contribute to symptomatology, since enations were observed on plants inoculated with either I17F or R1R2I3.

The second group of hosts corresponds to *Nicotiana clevelandii* and *Nicotiana glutinosa*, for which the symptoms are mainly determined by RNA 3. On *N. clevelandii*, R1R2I3 and R1R2RI3 are as severe as I17F, whereas the R strain induced only mosaic and weaker stunting. Thus, symptom severity on *N. clevelandii* is clearly related to the 3' part of the RNA 3 molecule. In *N. glutinosa*, both parental strains induced strong stunting, somewhat more severe with I17F than with R, whereas R–CMV in addition is responsible for pronounced leaf blistering. Since blistering was observed on plants inoculated with R1R2IR3, but was absent in those inoculated with R1R2I3, this symptom was also determined by the 3' part of the molecule. Thus, on both hosts, the severe symptoms determined by RNA 3 are due to the portion of the genome that includes the CP gene. Moreover, on *N. glutinosa*, since R1R2IR3 induced stunting more severe than that induced by the parental R strain, this suggests that in this case the 5'–half of RNA 3 also contributes to the severity of stunting.

Biological properties of CMV/TAV pseudorecombinant and recombinant strains
The replication properties of the different strains were tested in tobacco protoplasts. In this system it is possible to investigate the viability of strains including the newly created mutant (R3/NdeI, T3/NdeI) and chimeric RNAs (RT3, TR3) separately

Table 1. Symptoms induced by R and I17F-CMV and by pseudorecombinant and recombinant strains.

Strain / Host	I17F	R1R2R3	R1R2I3	R1R2RI3	R1R2IR3
Physalis floridana	mosaic, stunting, filiformism	mosaic	mosaic	mosaic	mosaic
Petunia hybrida	strong mosaic and stunting	mosaic, recovery	mosaic	mosaic	mosaic
Lycopersicon esculentum	mosaic, stunting, strong filiformism	almost no symptoms	mosaic, faint filiformism	mosaic, faint filiformism	mosaic, faint filiformism
Nicotiana glauca	mosaic, enation, stunting, filiformism	mosaic, recovery	mosaic, enation	mosaic	mosaic
Nicotiana tabacum cv. Xanthi or Samsum[a]	mosaic, enation, reduced area of leaves, stunting	mosaic, recovery	mosaic	mosaic	mosaic
Nicotiana clevelandii	Mosaic, strong stunting, filiformism	mosaic, weak stunting	mosaic, strong stunting, filiformism	mosaic, strong stunting, filiformism	mosaic, stunting
Nicotiana glutinosa	mosaic, strong stunting, filiformism	mosaic, blistering, strong stunting	mosaic	mosaic	mosaic, blistering, extreme stunting

[a] On tobacco, the I17F strain induces a mosaic that appears first as a general leaf chlorosis. Later, all or part of certain leaves are uniformally clearer, of reduced area and exhibit green enations. By contrast, all other strains induce a mosaic corresponding to numerous chlorotic spots.

from more complex functions (i.e. cell-to-cell and long-distance movement). Tobacco protoplasts were transfected with R1R2 transcripts together with R3/NdeI, T3/NdeI, RT3 or TR3 transcripts. The Northern analysis of the total RNA extracts showed that all four RNA3 molecules replicated to similar levels 24 hours after transfection, and the subgenomic RNA4, which encodes the coat protein, was able to be synthesized from RNA3 in all cases (Fig. 2).

Tobacco and *Chenopodium quinoa* plants were infected with the two parental strains (P-TAV, R-CMV) and also with the pseudorecombinant (R1R2T3, T1T2R3) and chimeric recombinant (R1R2RT3, R1R2TR3) strains which were tested in the protoplast experiment. All strains, except the chimeric R1R2TR3, infected tobacco systemically, and gave rise to local lesions on the leaves of *Chenopodium quinoa*. We repeated these inoculations several times, even with the same inoculum that was used for protoplast experiments, but were unable to detect any sign of infection with R1R2TR3, even in the inoculated leaves.

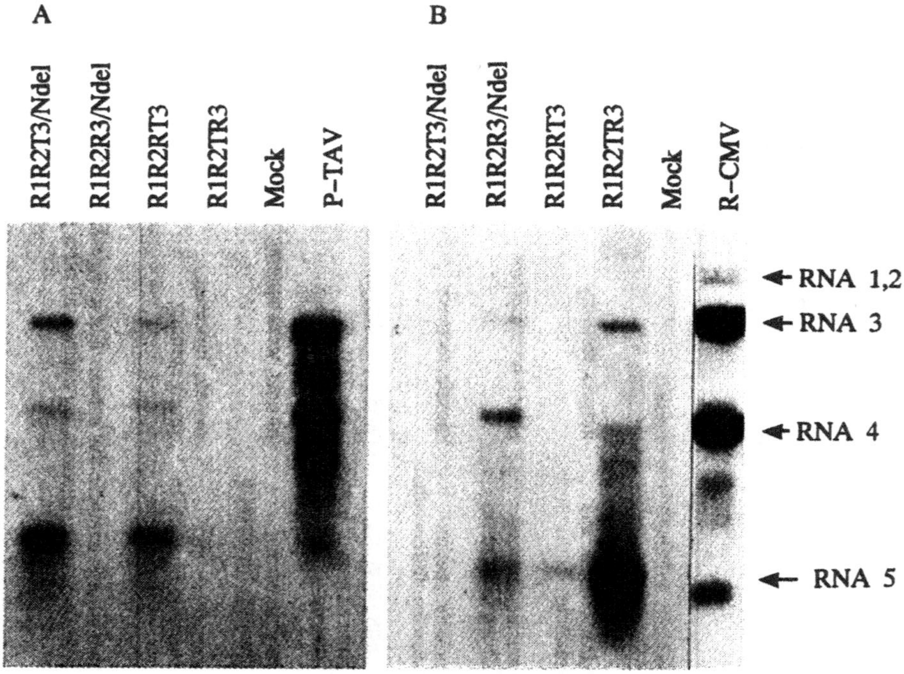

Figure 2. Northern blot analysis of progeny viral RNA in total RNAs of tobacco protoplasts. Protoplasts were transfected with *in vitro* transcripts from cDNA clones of R-CMV RNAs 1 and 2, together with transcripts from cDNA clones pT3/Nde, pR3/Nde, pTR and pRT. Protoplasts were also either mock-transfected, or transfected with viral RNAs purified from plants infected with P-TAV or R-CMV. Panel A shows results with a probe specific for the 3' end of TAV RNA 3, and panel B with one specific to the 3' end of CMV RNA 3. The position of RNAs 1, 2, 3 and 4 is indicated (Salánki *et al.*, 1997).

Table 2. Symptoms induced by P-TAV, R-CMV, pseudorecombinant and recombinant strains

Host	P-TAV	R-CMV	T1T2R3/Nde	R1R2T3/Nde	R1R2RT3	R1R2TR3
N. tabacum	mosaic[a]	mosaic	mosaic	mosaic	mosaic	-[b]
N. glutinosa	mosaic	mosaic, blistering stunting	mosaic, blistering stunting	mosaic	mosaic	-
N. benthamiana	mosaic, distortion stunting	mosaic stunting	mosaic strong stunting	mosaic stunting	mosaic stunting	-

[a] Only systemic symptoms are shown for the three *Nicotiana* species. All in vitro transcripts were able to replicate in *N. tabacum* protoplasts.
[b] - , absence of detectable infection.

Northern analysis confirmed this observation, as well as the conformity of the other strains (Salánki *et al.,* 1997). The different viruses were purified from infected tobacco plants, and the viral RNAs were extracted from the purified viruses. This viral RNA was used as inoculum in the following inoculation assays. The results of the different plant inoculation experiments are summarized in Table 2.

N. benthamiana plants were inoculated with the different purified viral RNAs. Both R-CMV and P-TAV induce mild mosaic and moderate stunting on this host. The R1R2T3 and the R1R2TR3 strains caused very similar symptoms to the parental strains, while T1T2R3 induced much more severe stunting than any of the other strains (Fig. 3). Northern analysis of the systemically infected *N. benthamiana* leaves showed that the differences in symptom gravity do not correlate with differences in level of virus accumulation, (Fig. 4), so the strong stunting in the case of T1T2R3 was not due to increased virus replication.

Figure 3. Typical symptoms elicited on *N. benthamiana* by, from left to right: R-CMV, P-TAV, R1R2T3/Nde pseudorecombinant virus, T1T2R3 pseudorecombinant virus, R1R2RT3 recombinant virus, and mock inoculation (3 weeks after inoculation).

On *N. glutinosa*, P-TAV induced a pronounced mosaic, while R-CMV induced not only mosaic but also severe stunting and leaf blistering (Fig. 5). The R1R2T3 pseudorecombinant virus induced symptoms equivalent to those of R-CMV. This shows that in this case stunting is determined by RNA 3. Considering the symptoms caused by the R1R2RT3 chimeric virus, mild mosaic and no stunting, this confirms, as was shown with the CMV recombinant strains above, that the major determinant of stunting induce by the R strain of CMV on *N. glutinosa* is within the 3' part of RNA 3.

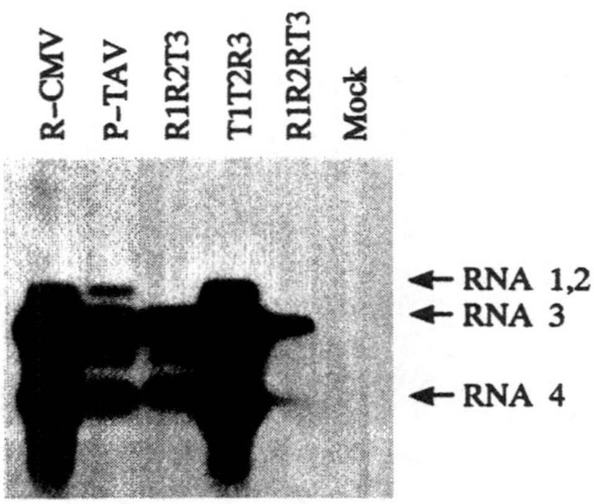

Figure 4. Nothern blot analysis of progeny viral RNA in total RNAs of infected *N. benthamiana* plants. The blot was probed with a mixed probe composed of clones corresponding to the 3' ends of TAV and CMV RNA 3. The position of RNAs 1, 2, 3 and 4 is indicated.

Discussion

Genetic mapping for symptomatology

There is growing evidence that determinants for symptomatology in cucumovirus-infected plants are very complex. The simplest case is when a typical symptom depends only on the expression of one gene - or on interaction between the host and a single part of the viral genome. This seems to be the case of the symptom of etch, which appears on the inoculated leaves of tobacco infected by CMV subgroup II strains, and which has been associated with the 3a gene and part of the flanking 5'-non-coding region (Zhang *et al.,* 1994). This seems also to be the case of leaf blistering in *N. glutinosa* infected by R-CMV, which is clearly dependent on the 3' part of RNA 3. The situation concerning *N. clevelandii* is less clear. The R1R2I3 strain induces symptoms similar to those of I17F, showing that RNA 3 is directly involved in symptomatology. Nevertheless, interpretation of the results obtained with chimeric RNAs 3 becomes more difficult, because we have to introduce a notion of disease severity. If indeed the recombinant containing the 3' part of I17F RNA 3 induces very strong symptoms, the converse recombinant, R1R2IR3 leads to an intermediate situation. The situation with *N. clevelandii* probably results from the fact that the disease phenotype includes various symptoms of different determinism: mosaic, filiformism and stunting.

Filiformism is induced on nearly all hosts by I17F-CMV. Although this symptom is more generally associated with infection of tomato, it has been clearly

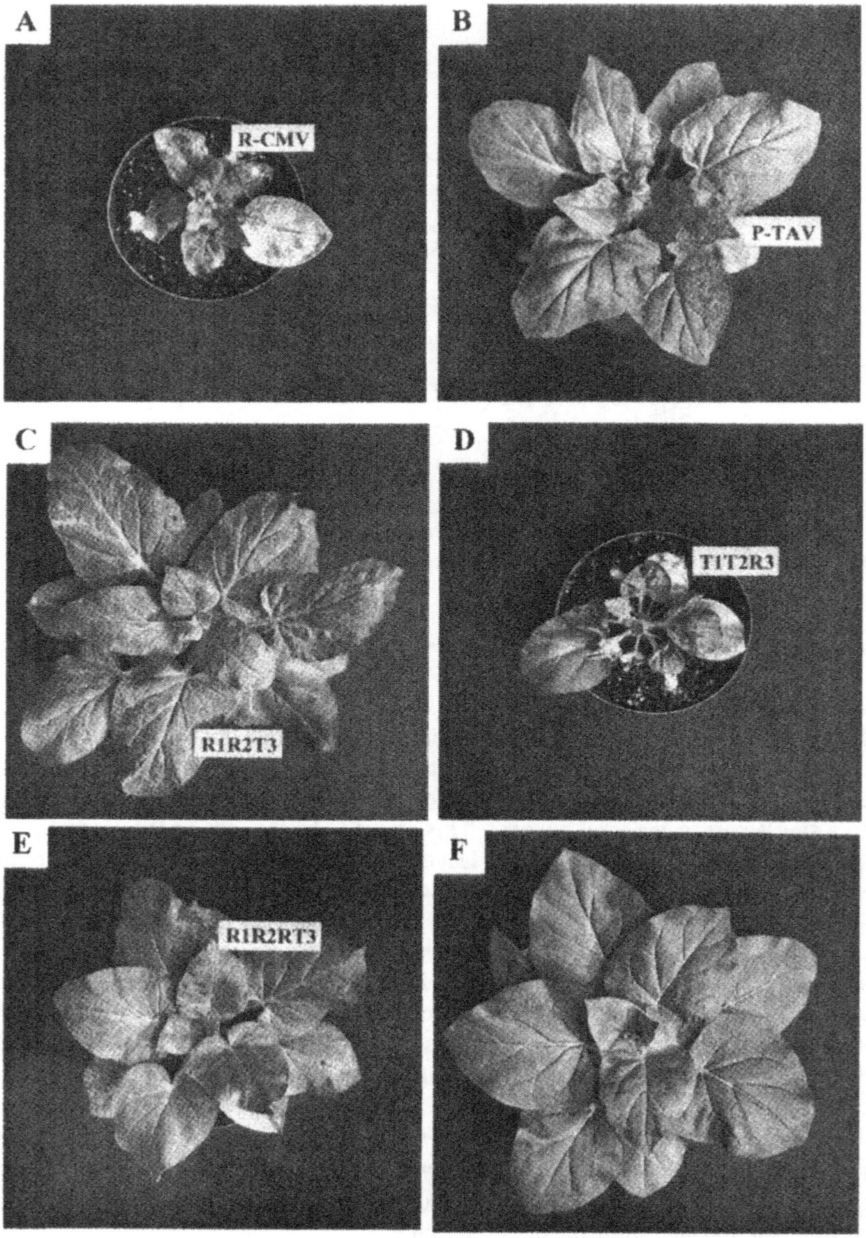

Figure 5. Typical symptoms elicited on *N. glutinosa* by (A) R–CMV, (B) P–TAV, (C) R1R2T3/Nde pseudorecombinant virus, (D) T1T2R3 pseudorecombinant virus, (E) R1R2RT3 recombinant virus, and (F) mock inoculation (3 weeks after inoculation).

observed in *N. glauca* and *N. glutinosa*, with one or two leaves reduced to their central vein. Since this symptom has been observed in tomato infected with other strains in this work, it is evident that filiformism cannot be simply associated with I17F RNAs 1 and/or 2. However, the symptoms were never as severe as with I17F.

Stunting is also probably the result of multiple determinants. For *P. floridana*, and to a lesser extent tomato and *N. glauca*, only I17F induces severe stunting. But for these hosts the main differences lie in the severity of the disease, rather than in the type of symptoms that develop. On *N. glutinosa*, stunting has been associated with the subgenomic RNA 4b (Ding *et al.*, 1995). Since both parental CMV strains, R and I17F, can be supposed to include a functional 2b gene, RNA 2 could contribute to the observed stunting. That RNA 3 is also involved is shown from both the recombinants between R and I17F CMV strains, and also those between R–CMV and P–TAV.

On tobacco the situation is made more complex by the phenomenon of alternate periods of symptom expression and disappearance, which correlate with the virus content in the corresponding leaves. This has been associated with the 3a movement protein (Gal–on *et al.*, 1996). Both R and I17F strains are subject to these cycles, but they are more pronounced for R–CMV, which cannot be recovered from certain leaves of infected plants. Since a similar pattern is observed for all strains that include RNAs 1 and 2 of R–CMV, since all CMV strains accumulated efficiently in the inoculated leaves, and since R is the strain for which this phenomenon is the most pronounced, this suggests that the deficiency of the R strain for long distance movement involves all three viral RNAs.

Movement of cucumoviruses in infected plants

The R1R2TR3 chimeric virus does not infect tobacco plants, even though this is a common host of CMV and TAV. This recombinant failed to move from cell–to–cell even thought it replicated in protoplasts. This is unexpected, since viable recombinants can be formed between much more distantly related viruses; for example, the movement protein (MP) and CP genes can be exchanged between bromo– and tobamoviruses to form viable viruses (De Jong and Ahlquist 1992). However, for these viruses a functional CP is not necessary for cell–to–cell movement, which is not the case in cucumovirus infection. CMV mutants bearing deletions in either the MP or in the CP gene are defective for cell–to–cell movement (Boccard and Baulcombe, 1993). Recently it was shown that the C–terminal 33 amino acids of the CMV MP are involved in the virus specificity of cell–to–cell movement (Nagano *et al.*, 1997), and that the 2a protein may also play a role in cell–to–cell movement (Nguyen *et al.*, 1996). So probably the efficient spread of cucumoviruses from the initially infected cells may require specific interactions between or among several cucumoviral proteins (2a, MP, CP) and/or the viral RNA itself. In this case the inability of the R1R2TR3 strain to infect cells could be the consequence of the improper interaction between the MP of P–TAV and the CP of R–CMV or the recombinant RNA itself. In any case, if this type of chimeric strain arises by recombination in transgenic plants, its multiplication will be limited to single cells, and so it will not have significant ecological impact.

Recombination, disease severity, and biosafety

Among the seven pseudorecombinant or recombinant cucumovirus strains we have created, two have induced symptoms worse than those of the parental strains on at least one host species. This frequency is worrisome, particularly since additional data from the litterature confirm that this is not rare. For instance, the natural pseudorecombinant between CMV and PSV recovered from common bean induces much more severe symptoms in this host than other cucumoviruses (White *et al.*, 1995). In addition, a recombinant created *de novo* by replacing the 2b gene of CMV by that of TAV, induces aggravated symptoms and accumulates to a higher level in different hosts (Ding *et al.*, 1996).

A second parameter which has to be considered in recombination assays is the putative selective advantage of a recombinant molecule versus the parental ones. Emergence of a recombinant RNA 3 with a clear selective advantage has already been described in a pseudorecombinant cucumovirus composed of RNAs 1 and 2 of CMV and RNA 3 of TAV (Fernandez-Cuartero *et al.*, 1994). From our results, we can reasonably propose that at least one recombinant, R1R2TR3, would not survive if it occurred naturally, since it cannot go beyond infection of initially infected cells. It will be of interest to carry out competition assays with the other cucumoviral recombinant RNAs 3 described here.

References

Boccard F, Baulcombe DC (1993) Mutational analysis of cis-acting sequences and gene function in RNA 3 of cucumber mosaic virus. Virology 193:563–578

De Jong W, Ahlquist P (1992) A hybrid plant RNA virus made by transferring the noncapsid movement protein from a rod-shaped to an icosahedral virus is competent for systemic infection. Proc Natl Acad Sci USA 89:6808–6812

Ding SW, Li WX, Symons RH (1995) A novel naturally occuring hybrid gene encoded by a plant RNA virus facilitates long distance virus movement. EMBO J 14:5762–5772

Ding SW, Shi BJ, Li WX, Symons RH (1996) An interspecies hybrid RNA virus is significantly more virulent than either parental virus. Proc Natl Acad Sci USA 93:7470–7474

Carrère I (1993) Contribution à l'étude de l'ARN 3 du virus de la mosaïque du concombre: rôle dans la symptomatologie et le transport du virus. thesis. Université de Paris XI, Paris

Fernandez-Cuartero B, Burgyán J, Aranda MA, Salánki K, Moriones E, Garcia-Arenal, F (1994) Increase in the relative fitness of a plant virus RNA associated with its recombinant nature. Virology 203:373–377

Gal S, Pisan B, Hohn T, Grimsley N, Hohn B (1992) Agroinfection of transgenic plants leads to viable cauliflower mosaic virus by intermolecular recombination. Virology 187:525–533

Gal-On A, Kaplan IB, Palukaitis P (1996) Characterization of cucumber mosaic virus. II. Identification of movement protein sequences that influence its accumulation and systemic infection in tobacco. Virology 226:354–361

Greene AE, Allison RF (1994) Recombination between viral RNA and transgenic plant transcripts. Science 263:1423–1425

Greene AE, Allison RF (1996) Deletions in the 3' untranslated region of cowpea chlorotic mottle virus transgene reduce recovery of recombinant viruses in transgenic plants. Virology 225:231–234

Hayes RJ, Buck KW (1990) Infectious cucumber mosaic virus RNA transcribed in vitro from clones obtained from cDNA amplified using the polymerase chain reaction. J Gen Virol 71:2503–2508

Jacquemond M, Lot H (1981) L'ARN satellite du virus de la mosaïque du concombre. I. Comparaison de l'aptitude à induire la nécrose de la tomate d'ARN satellites isolés de plusieurs souches du virus. Agronomie 1:927–932

Lommel SA, Xiong Z (1991) Reconstitution of a functional red clover necrotic mosaic virus by recombinational rescue of cell–to–cell movement gene expressed in a transgenic plant. J Cell Biochem 15A:151

Martin RR, Keese PK, Young MJ, Waterhouse PM, Gerlach WL (1990) Evolution and molecular biology of luteoviruses. Annu Rev Phytopathol 28:341–363

Nagano H, Okuno T, Mise K, Furusawa I (1997) Deletion of the C–terminal 33 amino acids of cucumber mosaic virus movement protein enables a chimeric brome mosaic virus to move from cell to cell. J Virol 71:2270–2276

Nagy JI, Maliga P (1976) Callus induction and plant regeneration from mesophyll protoplasts of *Nicotiana sylvestris*. Z. Pflanzenphysiol 78:453–455

Nguyen L, Lucas WJ, Ding B, Zaitlin M (1996) Viral RNA trafficking is inhibited in replicase–mediated resistant transgenic tobacco plants. Proc Natl Acad Sci USA 93:12643–12647

Revers F, Le Gall O, Candresse T, Le Romancer M, Dunez J (1996) Frequent occurrence of recombinant potyvirus isolates. J Gen Virol 77:1953–1965

Salánki K, Balázs E, Burgyán J (1994) Nucleotide sequence and infectious *in vitro* transcripts of RNA 3 of tomato aspermy virus pepper isolate. Virus Res 33:281–289

Salánki K, Carrère I, Jacquemond M, Balázs E, Tepfer M (1997) Biological properties of pseudorecombinant and recombinant strains created with cucumber mosaic virus (CMV) and tomato aspermy virus (TAV). J Virol 71:3597–3602

Sambrook J, Fritsch FF, Maniatis T (1989) Molecular cloning: a laboratory manual. Cold Spring Harbor Laboratory, Cold Spring Harbor, NY

Saunders JA, SMith CR, Kaper JM (1989) Effects of electroporation pulse wave on the incorporation of viral RNA into tobacco protoplasts. BioTechniques 7:1124–1131

Schoelz JE, Wintermantel WM (1993) Expansion of viral host range through complementation and recombination in transgenic plants. Plant Cell 5:1669–1679

Simon AE, Bujarski JJ (1994) RNA–RNA recombination and evolution in virus–infected plants. Annu Rev Phytopathol 32:337–362

White JL, Kaper JM (1989) A simple method for detection of viral satellite RNAs in small tissue samples. J Virol Methods 23:83–94

White PS, Morale, F, Roossinck MJ (1995) Interspecific reassortment of genomic segments in the evolution of cucumoviruses. Virology 207:334–337

Wintermantel WM, Schoelz JE (1996) Isolation of recombinant viruses between cauliflower mosaic virus and a viral gene in transgenic plants under conditions of moderate selection pressure. Virology 223:156–164

Zhang L, Hanada K, Palukaitis P (1994) Mapping local and systemic symptom determinants of cucumber mosaic cucumovirus in tobacco. J Gen Virol 75:3185–3191

Recombination Between Cauliflower Mosaic Virus and Transgenic Plants that Contain CaMV Transgenes : Influence of Selection Pressure on Isolation of Recombinants

William M. Wintermantel[1,2], Loránt Király[1], June Bourque[1], and James E. Schoelz[1]
[1]Department of Plant Pathology, University of Missouri, Columbia, Missouri 65211, USA
[2]Present Address: Department of Plant Pathology, 334 Plant Science Bldg. Cornell University, Ithaca NY 14853, USA

Abstract
We have shown that some cauliflower mosaic caulimovirus (CaMV) strains are able to recombine with *Nicotiana bigelovii* plants that contain a CaMV transgene that determines systemic infection of solanaceous species. The transgenic *N. bigelovii* expressed gene VI of CaMV strain D4, a gene known to determine systemic infection in solanaceous species including *N. bigelovii*. Recombinants between CaMV and the transgene can be readily isolated under conditions of strong selection pressure, but surprisingly can also be isolated from transgenic plants under moderate selection pressure. To simulate conditions of strong selection pressure, we inoculated the transgenic *N. bigelovii* with CaMV strain CM1841, a strain that is unable to infect *N. bigelovii* systemically. Recombinants between CM1841 and the transgene were isolated from 30% of the plants after mechanical inoculation. When CM1841 was agroinoculated, a recombinant virus was isolated from 100% of the transgenic plants. To simulate conditions of moderate selection pressure, we inoculated transgenic plants with CaMV strain W260, a virus that is able to infect nontransformed *N. bigelovii* systemically. We hypothesized that a recombinant virus formed between W260 and the transgene would have little or no selective advantage over the wild-type W260 virus. However, we found evidence that W260 had recombined with the transgene in three of 23 transgenic *N. bigelovii* infected with W260. The observation that a recombinant virus could predominate over the wild-type W260 virus was unforeseen, but might reflect differences in the aggressiveness of the recombinant relative to the initial inoculum. To determine whether the W260 recombinant virus, designated W260R, was more competitive than the wild-type W260, the two viruses were inoculated separately and in varying ratios onto non-transformed *N. bigelovii*. An analysis of the viruses responsible for the infections revealed that W260R could predominate in mixed infections, even when it was only a minor component of the initial virus population.

Introduction

It has recently been demonstrated that plant viruses can acquire an assortment of viral genes from transgenic plants. Lommel and Xiong (1991) showed that a red clover necrotic mosaic dianthovirus isolate defective in cell–to–cell movement recombined with a copy of the wild type cell–to–cell movement gene present in transgenic *Nicotiana benthamiana* plants. Cauliflower mosaic caulimovirus (CaMV) was also shown to be capable of acquiring a copy of its gene VI, a gene identified as a translational transactivator, from either transgenic *Brassica napus* (Gal *et al.*, 1992) or transgenic *Nicotiana bigelovii* (Schoelz and Wintermantel, 1993). Greene and Allison (1994) demonstrated that a cowpea chlorotic mottle bromovirus (CCMV) isolate that contained a defective coat protein gene could acquire a functional copy through recombination with transgenic *N. benthamiana* plants that contained CCMV coat protein sequences.

One feature common to the above studies is that strong selection pressure was used for the isolation of the recombinant virus from the transgenic plants. Strong selection pressure can be defined as the inability of a virus to infect a host systemically unless a recombination event occurs (Schoelz and Wintermantel, 1993; AIBS report, 1995). Experiments conducted under strong selection pressure have been significant for establishing that plant viruses do in fact recombine with viral transgenes, but they may be less likely to be representative of what would occur when transgenic plants are deployed in the field than experiments conducted under low selection pressure. Low selection pressure is defined as a situation in which the recombinant virus would have no selective advantage over the wild type virus.

We have observed that CaMV can readily acquire CaMV transgenes under conditions of strong selection pressure. In some experiments, recombinants can be isolated from 100% of the transgenic plants inoculated with CaMV. In addition, our studies now illustrate that recombinant viruses can be isolated from transgenic plants even when the initial virus can systemically infect the host, conditions that we describe as moderate selection pressure.

Results and Discussion

The experimental system for studying recombination between CaMV and CaMV transgenes.

CaMV is the type member of the caulimovirus group, a group whose genome consists of circular dsDNA approximately 8000 bp in length. A general map of CaMV showing the positions of the six major open coding regions and the functions of their protein products is depicted in Fig. 1. Previously, we identified the genes of CaMV that determine systemic infection of solanaceous hosts (Schoelz, *et al.*, 1986; Qiu and Schoelz, 1992). The CaMV model that evolved was based on six hosts and three CaMV strains as listed in Table 1. All three CaMV strains are known to infect the same crucifers, such as turnip, cauliflower and *B. napus*. They can be distinguished from one another by their effective response on the five solanaceous species listed in Table 1. CaMV strain CM1841 is unable to systemically infect any member of the *solanaceae* after mechanical inoculation. In contrast, strain D4 systemically infects all five solanaceous species, inducing systemic symptoms 18–24 days after inoculation. The host specificity of W260 lies

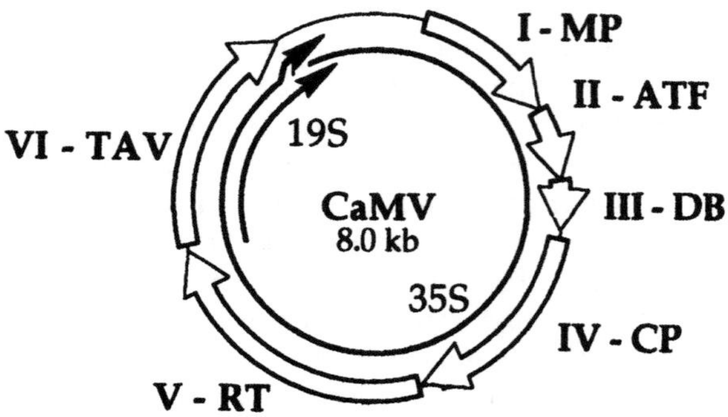

Fig. 1. Genome Structure of CaMV.
The positions of the six major genes of CaMV that comprise the circular genome are indicated by the outer arrows. The functions of the six genes are: MP, movement protein; ATF, aphid transmission factor; DB, DNA binding protein; CP, coat protein; RT, reverse transcriptase; and TAV, translational transactivator. The two curved arrows inside of the genome map indicate the positions of the two CaMV RNA transcripts, the 19S and 35S RNAs.

intermediate between D4 and CM1841, as W260 induces a systemic mosaic in *N. bigelovii*, a mixture of systemic mosaic and necrosis in *N. clevelandii*, systemic necrosis in *Datura stramonium*, and no systemic response in *N. edwardsonii* or *N. glutinosa*. In host–virus combinations that result in the lack of a systemic response, an enzyme–linked immunosorbant assay was conducted to verify that noninoculated leaves did not contain any virus. Thus, we can state with certainty that the absence of systemic symptoms is due to a block in systemic movement of the virus.

Table 1. Systemic Infection of Selected Solanaceous Species by Three CaMV Strains

Virus	*Nicotiana bigelovii*	*Nicotiana clevelandii*	*Nicotiana edwardsonii*	*Nicotiana glutinosa*	*Datura stramonium*
CM1841	None	None	None	None	None
W260	Systemic Mosaic	Systemic Necrosis	None	None	Systemic Necrosis
D4	Systemic Mosaic	Systemic Mosaic	Systemic Mosaic	Systemic Mosaic	Systemic Mosaic

By constructing chimeric viruses of W260 and D4 with CM1841, we have been able to identify genetic sequences of these two viruses that affect their ability to systemically infect solanaceous hosts. All of the chimeric viruses have been able to systemically infect turnips, the common host for all three CaMV strains, but they differ in their ability to infect the solanaceous species. The studies with chimeric viruses allowed us to conclude that gene VI of D4 determines systemic infection of solanaceous species, (Schoelz *et al.*, 1986), while W260 requires genes II, IV and VI to systemically infect *N. bigelovii* and *N. clevelandii* (Qiu and Schoelz, 1992; Wintermantel *et al.*, 1993; Qiu *et al.*, 1997).

Recombination between CaMV and CaMV transgenes under conditions of strong selection pressure.

We have demonstrated that CaMV can recombine with the CaMV gene VI coding region present in the genome of transgenic *N. bigelovii* (Schoelz and Wintermantel, 1993). In our studies, strong selective pressure was conferred by a difference in the host range determinants present in the viral inoculum and the host range determinants present within the viral transgene. Transgenic *N. bigelovii* expressed the gene VI product of CaMV strain D4, the CaMV gene that determines systemic infection of solanaceous species including *N. bigelovii*, while sequences flanking gene VI were derived from CM1841 (Fig. 2). Because the CM1841 wild-type virus is unable to infect *N. bigelovii* systemically, it was postulated that a CM1841 recombinant virus that would acquire the D4 transgene would gain the ability to systemically infect *N. bigelovii*, and therefore would exhibit a selective advantage over the wild-type CM1841.

The recovery of recombinants between CM1841 and the D4 transgene was shown to be extremely high. All plants inoculated with CM1841 wild-type virions that became infected systemically, approximately 30%, contained a recombinant virus (Table 2) (Wintermantel and Schoelz, 1996). When CM1841 was agroinoculated, a recombinant virus was isolated from 100% of the transgenic plants. The recombination event spans a substantial portion of the CaMV genome. To date is has been determined that the recombination event can involve up to 1928 bp, a DNA segment that includes the entire gene VI coding region (Schoelz and Wintermantel, 1993). DNA sequencing techniques have been used to map the recombination junctions of eight recombinant viruses isolated. Recombination usually occurred within DNA sequences of approximately 100 bp in length at the 5' and 3' ends of the gene VI mRNA produced by the transgenic plants (Fig. 2). In essence, very short regions of homology are sufficient for recombination to occur between wild type CaMV sequences and CaMV transgenes expressed in the host plant.

The primary mode of recombination probably consists of a RNA-RNA mechanism, involving a template switch during reverse transcription from the CaMV 35S RNA of the virus to the 19S RNA produced by the transgenic plants. The DNA segment introduced into the transgenic plants contained the CaMV 19S RNA promoter and polyadenylation signals in addition to the gene VI coding sequences (Schoelz and Wintermantel, 1993). Thus, the transgenic plants produce an authentic CaMV 19S RNA transcript. Recombination occurs during viral replication in the infected transgenic plants. During synthesis of the first strand of viral CaMV DNA, the reverse transcriptase switches templates, from the 5' end of the 35S RNA to the 3' end of the 35S RNA (Mason *et al.*, 1987). In a study of recombination between CaMV strains in turnips, Vaden and Melcher (1990) suggested that when more than one strain is inoculated, the 19S RNA might be reverse transcribed into DNA through a template switch from the 5' end of the 35S RNA to the 3' end of the 19S RNA of another strain. The 35S and 19S RNAs of CaMV have the same 3' terminus (Covey *et al.*, 1981; Guilley *et al.*, 1982), which might explain how a template switch could occur. Vaden and Melcher (1990)

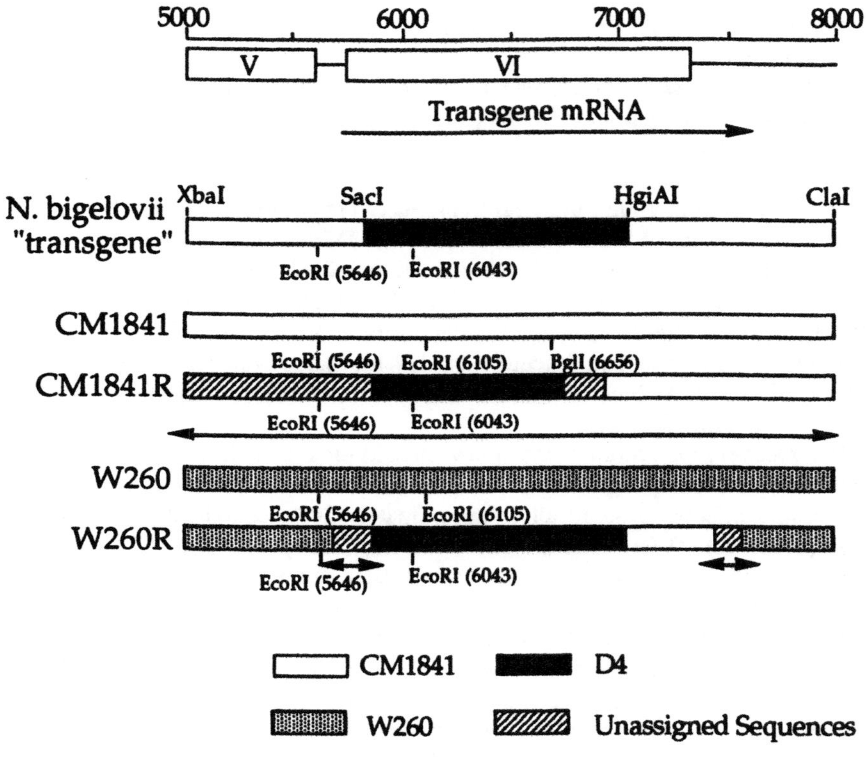

Fig. 2. Location of recombination junctions between CaMV and transgene sequences.
Partial EcoRI maps of CM1841, W260, and CaMV transgene sequences. Transgenic plants contained a *Xba*I – *Cla*I DNA segment of CaMV and produced a mRNA from the CaMV 19S promoter and polyadenylation signals. A *Sac*I – *Hgi*AI DNA segment of the transgene was derived from CaMV strain D4. This segment contained the essential D4 sequences necessary for systemic infection of solanaceous species. Regions that flank the transgene were derived from strain CM1841. After inoculation with either CM1841 or W260, virion DNAs were recovered from transgenic plants by PCR or by passage through turnips. The double arrows indicate the regions that were sequenced to identify the recombination junctions. The unassigned sequences in the diagram cannot be identified as being derived from the transgene or the CaMV viral inoculum because of a lack of sequence polymorphisms. In the case of CM1841R, approximately 3000 bp were sequenced to determine the extent of recombination between CM1841 and the D4 transgene. There are 39 nucleotide differences between the D4 transgene sequence and gene VI of CM1841. The sequence of CM1841R agreed with the transgene sequence at 38 of those sites, indicating that the recombination event spanned at least 856 nucleotides, as indicated. In the case of W260R, recombination with the transgene was confirmed by localization of the recombination junctions and the presence of an *Eco*RI site at nucleotide position 6043, a site not found in wild type W260.

Table 2. Isolation of Recombinant Viruses from Transgenic *N. bigelovii* Plants

# Plants with systemic symptoms[a]				# Infections analyzed[b]			
Virus	Number of Replications	Wild-type *N. bigelovii*	Transgenic *N. bigelovii*	W	R	M	Total Analyzed
CM1841	13	0/102	38/106	0	22	2	24
W260	12	101/102	104/106	20	1	2	23

[a]The total number of plants which developed systemic symptoms / the number of plants inoculated.
[b]Number of infected plants analyzed in which the systemic infection resulted from: (W) the wild type virus; (R) recombination with the transgene; (M) a mixture of wild type and recombinant viruses.

suggested that when the reverse transcriptase reached the 5' end of the 19S RNA, it would switch back to the 35S RNA to complete the synthesis of the first strand of CaMV DNA. Our nucleotide sequence information suggests that a similar template switch has occurred between the 35S RNA of the wild type viral CM1841 and the 19S RNA produced by the transgenic plants. Although our evidence indicates that template switching may be responsible for recombination events, it is also possible that homologous recombination may occur between the virus and the transgene sequences. Gal *et al.*, (1992) observed that one recombinant virus (out of seven analyzed) may have arisen through homologous recombination. It is possible that CaMV is capable of recombination by both mechanisms.

To study the temporal and spatial appearance of recombinant viruses in transgenic plants, we inoculated CM1841 by agroinoculation and by mechanical inoculation of virions. In agroinoculated transgenic plants, inoculated leaf tissue turned necrotic in the area surrounding the site of inoculation. In virion-inoculated plants, local lesions were chlorotic. However, the first transgenic *N. bigelovii* leaves to develop systemic symptoms became necrotic, regardless of whether they were agroinoculated or inoculated with virions. The predominant type of virus recovered from the inoculated and first systemically infected leaves was the wild type CM1841 rather than a recombinant. As the infection progressed up the transgenic plants, symptoms changed from necrosis in the lower leaves to a chlorotic mosaic in the upper leaves. This shift in symptom type was associated with the increase of recovery of recombinant viruses in upper leaves in both agroinoculated and virion-inoculated plants, indicating that recombinants predominated only in later stages of pathogenesis.

The recombinant viruses isolated from the transgenic plants, as characterized by recombinant virus CM1841R, differ phenotypically from the wild-type CM1841 in two ways. CM1841R induces symptomless infections in turnips, differing from wild type CM1841, which causes a distinct chlorotic mosaic. Second, CM1841R induced a systemic mosaic in non-transformed *N. bigelovii*, *N. glutinosa*, *N. clevelandii*, *N. edwardsonii* and *D. stramonium* (Schoelz and Wintermantel, 1993). In contrast, CM1841 infections are limited to the inoculated leaves of these plants. Both of these phenotypic traits could be predicted from the gene VI coding sequence of CM1841R. Previous genetic studies showed that chimeric virus H7 (Daubert *et al.*, 1984), a virus constructed *in vitro* from D4 and CM1841 that has a genome composition similar to CM1841R, also gave rise to symptomless infections in turnips, but a systemic mosaic *in D. stramonium*, *N. bigelovii* and *N. edwardsonii* (Daubert *et al.*, 1984; Schoelz *et al.*, 1986; Schoelz and Shepherd, 1988).

Although the recombinant virus CM1841R caused attenuated symptoms in turnips, it is possible that recombination between CaMV and transgene sequences could result in a virus infection that is more severe than that brought on by the virus that was originally inoculated. The overall effect of recombination on symptomatology in a host may be predicted, to a certain extent, by the construction of chimeric viruses *in vitro*. For example, the work of Anderson *et al.* (1992) predicts that recombination between genomic sequences of a transgenic plant that contains the CM1841 gene VI coding sequence and viral sequences of CaMV strain W260 could potentially result in a strain of CaMV that is significantly more severe in turnips than the wild type W260 virus. Similarly, alterations in host specificity might be foreseen by an understanding of the role the viral transgene plays in influencing host specificity. The uncertainty in calculating the effects of recombination on symptomatology may become critical when an uncharacterized strain infects a transgenic plant.

Recombination between CaMV and CaMV transgenes under conditions of moderate selection pressure.

It had been previously thought that strong selection pressure may be necessary to recover recombinant viruses from transgenic plants (Schoelz and Wintermantel, 1993; Falk and Bruening, 1994). The implication was that recombination might occur in the absence of selection pressure, but in the absence of a selective advantage the recombinant virus would be unable to move in the infected plant due to exclusion by the wild-type virus. We have now demonstrated, however, that recombinant viruses between the gene VI transgene and CaMV can be isolated from transgenic plants even when the initial virus can systemically infect the host, conditions that can be described as moderate to weakly selective.

To determine whether recombinant viruses could be recovered from transgenic plants under conditions of moderate selection pressure, we examined the infections of transgenic *N. bigelovii* caused by CaMV strain W260, a virus that infects non-transformed *N. bigelovii* systemically (Schoelz and Shepherd, 1988). Hence, a recombinant virus formed between W260 and the transgene would presumably have little selective advantage over the initial viral inoculum. Virus recovered from systemically infected leaves of 23 transgenic *N. bigelovii* established that 20 plants had developed infections resulting from the systemic movement of the wild-type W260 virus, while a recombinant virus was responsible for symptoms in the three remaining plants (Table 2). Two of these latter three infections consisted of a mixed population of wild type W260 and a recombinant. In one plant, only the recombinant virus was recovered from systemically infected leaves, indicating that the recombination event occurred early after the initial infection and had successfully excluded the movement of the wild type W260 into the upper leaves of the host. Characterization of the sequences of viral DNA recovered from the infection in which only the recombinant was detected revealed that the recombination junctions were localized to short stretches of homologous sequences that contained both the transgene transcript initiation and termination sites (Fig. 2). This provides further evidence that the mechanism of recombination involves template switches between the 35S RNA and the transgene mRNA during reverse transcription (Gal *et al.,* 1992; Schoelz and Wintermantel, 1993). The W260 recombinant virus arose when the reverse transcriptase switched from the 35S RNA template to the transgene mRNA in a homologous region within 67 nucleotides from the 3' end of the transgene mRNA. A second crossover, from the transgene mRNA back to the viral 35S RNA transcript, was localized within 73 nucleotides of the 5' end of the transgene mRNA. In essence, the entire gene VI coding region of the W260 viral sequence had been replaced with the corresponding sequence from the transgene (Wintermantel and Schoelz, 1996).

The observation that a recombinant virus could predominate over the wild type W260 virus may perhaps reflect differences in the aggressiveness of the recombinant virus relative to the initial inoculum. For example, the host specificity information present on the transgene, which is derived from CaMV strain D4, may confer a selective advantage to a recombinant virus over wild-type or other recombinant viruses in which gene VI is derived from W260. Although both D4 and W260 are capable of infecting *N. bigelovii* systemically, D4 systemic symptoms usually appear at 18 – 20 days postinoculation, 4 – 5 days before those caused from a W260 infection (Schoelz and Shepherd, 1988).

To determine whether the W260 recombinant virus, designated W260R, was in actuality more competitive than the wild type W260, the two viruses were inoculated onto non-transformed *N. bigelovii* separately and in ratios of 1:1, 5:1, and 1:5 (W260 vs. W260R, total virion concentration of 75 μg/ml) (Wintermantel and Schoelz, 1996). The infections induced by W260 and W260R were indistinguishable until systemic symptoms developed. There was no difference noted in the temporal appearance of primary lesions induced by W260 or W260R and similarly, no difference between the specific infectivity of the two virion preparations. In contrast to the local lesions, systemic symptoms caused by W260R began 4 days before those of W260, indicating that W260R was more aggressive than W260. In addition, the *N. bigelovii* plants inoculated with mixtures of W260 and W260R developed systemic symptoms at the same rate as those inoculated with W260R alone, further suggesting that W260R could be responsible for the shorter time needed for systemic infection when compared to the wild type alone. An analysis of the virus responsible for the systemic infection in plants inoculated with mixtures of the W260 and W260R revealed that W260R could predominate in mixed infections even when comprising only a minor component of the initial virus population. W260R was the only virus recovered from each of the 10 plants inoculated at ratios of 5:1 or 1:1 (W260R vs. W260). Of the plants inoculated at a ratio of 1:5 (W260R vs. W260), W260R alone was isolated from 9 plants, while W260 alone was recovered in one plant.

This study has demonstrated that recombination between a virus and a transgene containing viral sequences can actually enhance the competitiveness of a virus if the transgene confers even a slight selective advantage and recombination occurs between closely related strains. It is important to note, however, that a selective advantage in one host may not mean that the virus has a selective advantage in all hosts. Significantly, mixtures of wild type and recombinant viruses detected in the transgenic plants remained mixtures even after passage through turnips, an indication that the recombinant may not have a selective advantage over the wild type in all hosts.

Conclusions

There are at least two approaches to investigate whether plants carrying viral transgenes pose a risk to the environment. The first involves carefully designed field studies to monitor transgenic plants for the appearance of recombinant viruses. Although these studies will be essential for demonstrating that transgenic plants do not present a risk to the environment, one potential problem with field studies concerns the size of the sample that would be required to prove that the release of transgenic organisms poses little or no risk. Critics of field studies might contend that it would not be possible to detect the rare recombinant, and that these recombinants might represent a biological risk. A second strategy that is worth pursuing in addition to field studies is to understand the factors that govern recombination between viruses and transgenes. If we understand how and when recombination occurs, we might be able to conclude that certain types (or perhaps all types) of pathogen–derived resistance are safe. This second approach may complement the field studies by providing an explanation for negative results (i.e. why recombinants are not detected in the field).

We have investigated the potential for recombination between sequences of CaMV and transgenic plants under different levels of selection pressure. In past years we have identified a single gene of CaMV, gene VI of strain D4, that determines systemic infection of solanaceous hosts, including the host *N. bigelovii* (Schoelz *et al.*, 1986; Schoelz and Shepherd, 1988). Recently we expressed the D4 gene VI product in transgenic *N. bigelovii* and demonstrated that under strong selection conditions, the recovery of CaMV recombinants from transgenic plants is very high. A high rate of recovery of recombinants in greenhouse studies is essential for evaluating the potential for recombination using different pathogen–derived resistance strategies. If, for example, recombinants are detected infrequently in the positive controls, it will be difficult to demonstrate experimentally that some pathogen–derived resistance strategies can preclude the formation of recombinants. We have also found that recombinant viruses between the gene VI transgene and CaMV can be isolated from transgenic plants under moderately selective conditions. In many field situations it is likely that the recombinant virus will have no selective advantage over the wild–type in the transgenic plant. The observation that CaMV recombinants can be isolated under conditions of moderate selection pressure is valuable because it will enable us to more precisely model the conditions that might occur with pathogen-derived resistance.

Although CaMV is classified as a DNA virus, our results also have relevance for RNA viruses because the mechanism of recombination between CaMV and transgenic plants is similar to the recombination mechanism between RNA viruses and transgenic plants. The primary mechanism of CaMV recombination involves a template switch during reverse transcription of the CaMV 35S RNA to the transgene mRNA produced by the transgenic plants. This is a RNA-RNA recombination mechanism (Gal *et al.,* 1992; Schoelz and Wintermantel, 1993). Recombination between RNA viruses involves a copy choice mechanism in which the RNA polymerase switches from one template to another location on the template or to a different template (Cascone *et al.,* 1993). Although the enzyme responsible for synthesis of viral nucleic acid differs between CaMV and RNA viruses, the same recombination mechanism is utilized by RNA viruses and CaMV. Therefore the results of our studies may also be useful for applications involving transgenic plant resistance to RNA viruses.

References

American Institute of Biological Sciences (AIBS). (1995) Transgenic virus resistant plants and new plant viruses. Meeting report from AIBS workshop sponsored by U. S. Department of Agriculture. 47 pp

Anderson EJ, Trese, AT, Sehgal OP, Schoelz JE (1992) Characterization of a chimeric cauliflower mosaic virus isolate that is more severe and accumulates to higher concentrations than either of the strains from which it was derived. Molec Plant-Microbe Interact 5:48-54

Cascone PJ, Haydar TF, Simon AE (1993) Sequences and structures required for recombination between virus-associated RNAs. Science 260:801-805

Covey SN, Lomonossoff GP, Hull R (1981) Characterisation of cauliflower mosaic virus DNA sequences which encode major polyadenylated transcripts. Nucl Acids Res 9:6735-6747

Daubert SD, Schoelz J, DeBao L, Shepherd RJ (1984) Expression of disease symptoms in CaMV genomic hybrids. J Mol Appl Genet 2:537-547

Falk BW, Bruening G (1994) Will transgenic crops generate new viruses and new diseases? Science 263:1395-1396

Gal S, Pisan B, Hohn T, Grimsley N, Hohn B (1992) Agroinfection of transgenic plants leads to viable cauliflower mosaic virus by intermolecular recombination. Virology 187:525-533

Greene AE, Allison RF (1994) Recombination between viral RNA and transgenic plant transcripts. Science 263:1423-1425

Guilley H, Dudley RK, Jonard G, Balazs E, Richards KE (1982) Transcription of cauliflower mosaic virus DNA: detection of promoter sequences and characterization of transcripts. Cell 30:763-773

Lommel SA, Xiong Z (1991) Reconstitution of a functional red clover necrotic mosaic virus by recombinational rescue of the cell-to-cell movement gene expressed in a transgenic plant. J Cell Biochem 15A:151

Mason WS, Taylor JM, Hull R (1987) Retroid virus genome replication. Adv. Virus Res. 32:35-96

Qiu SG, Schoelz JE (1992) Three regions of cauliflower mosaic virus strain W260 are involved in systemic infection of solanaceous hosts. Virology 190:773-782

Qiu SG, Wintermantel WM, Sha Y, Schoelz JE (1997) Light-dependent systemic infection of solanaceous species by cauliflower mosaic virus can be conditioned by a viral gene encoding an aphid transmission factor. Virology 227:180–188

Schoelz JE, Shepherd RJ, Daubert SD (1986) Region VI of cauliflower mosaic virus encodes a host range determinant. Mol Cell Biol 6:2632–2637

Schoelz JE, Shepherd RJ (1988) Host range control of cauliflower mosaic virus. Virology 162:30–37

Schoelz JE, Wintermantel WM (1993) Expansion of viral host range through complementation and recombination in transgenic plants. Plant Cell 5:1669–1679

Vaden VR, Melcher U (1990) Recombination sites in cauliflower mosaic virus DNAs: Implications for mechanisms of recombination. Virology 177:717–726

Wintermantel WM, Anderson EJ, Schoelz JE (1993) Identification of domains within gene VI of cauliflower mosaic virus that influence systemic infection of *Nicotiana bigelovii* in a light-dependent manner. Virology 196:789–798

Wintermantel WM, Schoelz JE (1996) Isolation of recombinant viruses between cauliflower mosaic virus and a viral gene in transgenic plants under conditions of moderate selection pressure. Virology 223:156–164

Synergy of Virus Accumulation and Pathology in Transgenic Plants Expressing Viral Sequences

Peter Palukaitis[1,2] and Igor B. Kaplan[1]

[1] Dept. of Plant Pathology, Cornell University, Ithaca, NY 14853 USA

[2] Virology Department, Scottish Crop Research Institute, Invergowrie, Dundee, DD2 5DA, UK

Abstract

Some viruses interact with other viruses resulting in an increase in the titre of one or both of the unrelated viruses, and an enhancement of the pathogencity induced by either virus. This interaction is called synergism. For one well-know type of synergism, it has been shown that a transgene containing particular sequences of one virus can enhance the accumulation and pathogenicity induced by a second virus. However, those transgenic viral sequences were not among the ones that have been used to date in studies on pathogen-mediated resistance.

Transgenic plants expressing viral sequences that result in resistance to the viral source of the transgene, and closely related viruses have been generated in a number of laboratories. Such transgenic plants generally show no resistance to infection by unrelated viruses. However, the extent to which such transgenic plants shown synergism with unrelated viruses, leading to increased virus accumulation and/or enhanced pathogencity has not been explored. Using transgenic plants generated in previous studies on replicase-mediated resistance, or for complementation studies, we have examined the ability of those transgenic plants to enhance the accumulation and/or pathogenicity of an unrelated virus; i.e., to substitute for one of the viral genome pairs in several known examples of viral synergism. These include transgenic plants expressing (1) the genes for the tobacco mosaic tobamovirus (TMV) 126K, 54K and 30K proteins, infected with cucumber mosaic cucumovirus (CMV), potato potexvirus X (PVX), and potato potyvirus Y (PVY); (2) the genes for the CMV 1a, 2a, and 3a proteins, infected with PVX, PVY, and TMV; and (3) the PVY NIb gene, infected with CMV, PVX, and TMV. The results of these tests will be presented, as well as the effect of temperature on the synergism.

Introduction

Some plant viruses can interact with each other producing a worse disease than caused by either virus alone. This synergy in pathology also often results in an increased accumulation for one of the viruses involved. Since infection of plants by multiple viruses is common (Kassanis, 1963), there are considerable opportunities for synergistic interactions leading to enhanced pathology in field-infected plants. Some control strategies for virus disease utilize the transgenic expression of viral sequences, cloned for particular parts of viral genomes, to protect such plants from the pathological effects caused by the same or related viruses (reviewed by Hull and Davies, 1992; Scholthof *et al.*, 1993; Wilson, 1993; Lomonossoff, 1995). These strategies generally have not taken into account the possibility of synergistic interactions between the viral transgene and another virus, in which the latter virus can interact synergistically with the viral progenitor of the transgene. Recently, however, one study involving synergism between potato potexvirus X (PVX) and several potyviruses demonstrated that the transgenic expression of only part of the potyviral genome was sufficient to enhance the replication of PVX and induce a pathological synergism (Vance *et al.*, 1995). Thus, synergism can still occur in the absence of the replication of one of the viral partners, and with a much lower level of gene expression than is observed in an infected plant.

We have generated a number of transgenic tobacco plants that express different viral sequences. Some of these transgenic plants show resistance to the homologous virus and some do not (Golemboski *et al.*, 1990; Anderson *et al.*, 1992; Audy *et al.*, 1994; Kaplan *et al.*, 1995; T. Canto, W. Wintermantel, M. Zaitlin and P. Palukaitis, unpublished data). These transgenic plants express genes derived from tobacco mosaic tobamovirus (TMV: 30K, 54K, 126K), cucumber mosaic cucumovirus (CMV: 3a, full-length RNA-1, full-length RNA-2, RNA-2 containing a 94-nt frameshift deletion in the 2a gene), or potato potyvirus Y (PVY: NIb gene). To determine whether such plants could induce a synergistic interaction with viruses known to interact synergistically with TMV, CMV, or PVY, these transgenic plants were inoculated with the above viruses as well as PVX, which is also known to interact synergistically with the above viruses.

Other technologies for engendering virus resistance also exist. Among these are cross-protection, the pre-infection of plants with mild or symptomless strains of a severe virus, and the use of natural resistance genes, introgressed by plant breeding into agriculturally important cultivars from wild relatives. For one particular case of synergy, that of CMV and zucchini yellow mosaic potyvirus (ZYMV) in cucurbit hosts, we have assessed the risks of synergy associated with the application of such alternative strategies.

Results

Characterization of Synergy between Four Viruses

Since synergy can be affected by the strain of the virus used (Goodman and Ross, 1974), and possibly by host genetic factors and the environment (Close, 1964), we reexamined the nature of synergy between different combinations of CMV, PVX, PVY, and TMV (Table 1). As can be seen, synergy was observed between all combinations of the above four viruses. In most cases, there was also an increase in the level of CMV and PVX, but there was no increase observed in the titre of PVY or TMV. This was true even though PVY and TMV together showed synergy in pathology. In many cases, the synergy in pathology preceded the increase in virus titre.

Table 1. Biological effects of mixed infections in tobacco.

VIRUSES	CMV	PVY	TMV
CMV	sMo, St	vsMo, St, sLD C>16	vsMo, sSt, vsLD C>16
PVX	vsVc, sN X>16	sN X>8	vsMo, St, LD X>8
PVY	vsMo, St, sLD C>16	vmVc	vsMo, St, LD V=
TMV	vsMo, sSt, vsLD C>16	vsMo, St, LD V=	sMo, St

Mo = Mosaic; St = Stunting; LD = Leaf Distortion; N = Necrosis; Vc = Vein–clearing. vm = very mild; s = severe; vs = very severe.

V= means the virus levels (9 d.p.i.) did not change; X># means an increase in PVX level; C># means an increase in CMV level.

Detection of Synergy in Transgenic Tobacco Expressing Specific Viral Sequences

Most of the combinations of transgenic tobacco expressing CMV or TMV sequences did not show synergy after inoculation with CMV, PVX, PVY, or TMV, although certain combinations did show synergy (Table 2). However, most of the transgenic tobacco lines that showed synergy expressed functional copies of viral genes (e.g., the CMV–1a protein or the TMV–30K protein). It is not known if the homologous genes expressing dysfunctional versions of those encoded proteins would also show synergy, assuming that they also exhibit resistance to the homologous virus. In two cases, transgenic tobacco expressing resistance to CMV (D–CMV RNA–2) or PVY (PVY–NIb) displayed synergy with an unrelated virus (Table 2). In both cases, the synergy observed was limited to some intensification of the disease and there was no corresponding increase in virus accumulation (of either PVY or TMV). In those instances, the corresponding transgenic plants expressing versions of the same genes that did not show resistance to the homologous viruses also did not show synergy.

None of the transgenic lines showed synergy with PVX, even though PVX showed synergy with CMV, PVY, and TMV (Table 1). Synergy between PVX and tobacco etch potyvirus (TEV) was shown to be associated with sequences in the 5' region of TEV (Vance *et al.*, 1995). Therefore, the data here indicate that for synergy between PVX and CMV, or PVX and TMV, either regions of the genome not represented in any of the transgenic plant lines, or combinations of such sequences are required to affect synergy with PVX. Since the TMV coat protein (gene) does not interact synergistically with PVX (Anderson *et al.*, 1989), it seems more likely that synergy between PVX and TMV involves either some combination of viral sequences, or is due to some effect of TMV itself that leads to an increase in PVX accumulation.

Detection of Heterologous Virus Interference in Transgenic Tobacco Expressing Homologous–Virus Resistance

Surprisingly, some transgenic plants expressing sequences of one virus showed interference with an unrelated virus (Table 2). This was particularly true of transgenic tobacco expressing the 126K gene of TMV, which showed neither resistance to TMV nor synergy with any of the other viruses tested, but showed some reductions in virus accumulation after infection by CMV, PVX, or PVY (Table 2). Transgenic tobacco expressing the TMV–54K gene and showing resistance to TMV also showed some interference with the accumulation of PVY, but no effect on the mild symptoms induced by PVY, while trangenic tobacco expressing the genes encoding functional movement proteins of CMV (3a) and TMV (30K) showed both a reduction in virus accumulation and in the pathology induced by PVX (Table 2). By contrast, transgenic tobacco expressing a functional TMV–30K gene stimulated infection by CMV and TMV (Table 2) as well as infection by a number of other viruses (Cooper *et al.*, 1995). Transgenic plants expressing the NIb gene of PVY gave the most unexpected results: A line showing resistance to PVY showed synergy with TMV *and* interference with both the pathology and virus accumulation of PVX (Table 2). A similar contrast of interactions was also observed in transgenic tobacco expressing the TMV–30K gene (Table 2).

Synergy between CMV and ZYMV

CMV and ZYMV show synergy in cucumber (Poolpol and Inouye, 1986). This synergy has also been demonstrated in melon and squash (A. Gal–On, personal communication). Since CMV accumulation and CMV–induced disease in cucumber can be suppressed by the use of natural resistance genes, and ZYMV–induced disease can be controlled by crossprotection using a mild strain of ZYMV, we wanted to establish whether such classical control strategies also carried risks of synergy.

Cucumber plants of the cultivar 'Straight 8' were highly susceptible to infection by CMV. This infection was manifested by high virus accumulation in the inoculated cotyledons and even higher levels of accumulation of CMV in systemically–infected leaves (Table 3). In such susceptible cucumber plants inoculated with CMV + ZYMV, there was a 16–fold stimulation of CMV levels in the inoculated cotyledon observable by 5 days post inoculations (d.p.i.), but no increase in the CMV titre in the leaves systemically–infected by CMV + ZYMV (Table 3). However, at 10 d.p.i., by which time the pathological effects of the CMV + ZYMV synergy had become manifested, the level of CMV in the cucumber leaves systemically–infected by CMV + ZYMV was 8–fold higher than the CMV level in cucumber leaves infected by CMV alone. These increases in CMV titre and the pathological effects of the synergy were manifested, whether the ZYMV was a severe or a mild strain (A. Gal–On, personal communication).

Table 2. Interactions observed between CMV, PVX, PVY and TMV in transgenic tobacco expressing CMV, PVX, PVY and TMV sequences.

TRANSGENE	VIRUS	EFFECT
INTERFERENCE		
TMV–30K (S)	PVX	(Pathology & Virus Level)
TMV–54K (R)	PVY	(Virus Level)
TMV–126K (S)	PVX	(Virus Level)
TMV–126K (S)	PVY	(Virus Level)
TMV–126K (S)	CMV	(Virus Level)
CMV–3a (S)	PVX	(Pathology & Virus Level)
PVY–NIb (R)	PVX	(Pathology & Virus Level)
SYNERGY		
fl–CMV RNA–1 (S)	PVY	(Pathology & Virus Level)
fl–CMV RNA–1 (S)	TMV	(Pathology & Virus Level)
D–CMV RNA–2 (R)	PVY	(Pathology – minor)
TMV–30K (S)	CMV	(Pathology & Virus Level)
TMV–30K (S)	PVY	(Pathology & Virus Level)
PVY–NIb (R)	TMV	(Pathology)

(S) = susceptible and (R) = resistant to homologous virus.

fl = full–length RNA; D = RNA–2 containing an inframe 94–nt deletion in the 2a gene. Other transgenes express sequences of the corresponding viral–encoded protein.

When cucumber plants of the CMV–resistant cultivar 'Delila' were inoculated by CMV alone, there was some accumulation of CMV in the cotyledons but none detectable in the leaves at 5 d.p.i., although at 10 d.p.i., there was some virus accumulating in the first leaf above the cotyledon (Table 3). At 5 d.p.i., the relative accumulation of CMV in the CMV–resistant cultivar was 32– to 64–fold lower than in cotyledons of the susceptible cultivar, while at 10 d.p.i., the first systemically–infected leaf of the CMV–resistant plant contained 64– to 128–fold less virus than the comparable leaf of the susceptible cultivar.

Table 3. Relative CMV accumulation in cucumber susceptible or resistant to CMV after inoculation with CMV or CMV + ZYMV.

TISSUE	RESISTANT CULTIVAR		SUSCEPTIBLE CULTIVAR	
	CMV	CMV+ZYMV	CMV	CMV+ ZYMV
Cotyledon 5 d.p.i.	1	4–8	32–64	512–1024
Systemic Leaf 5 d.p.i.	0	32–64	512–1024	512–1024
Systemic Leaf 10 d.p.i.	1	64–128	64–128	512–1024

Relative levels of CMV, determined by ELISA with the A405 nm values for CMV alone in the resistant cultivar 'Delila' taken as 1, in the cotyledons at 5 d.p.i. and in the first leaf above the cotyledons at 10 d.p.i.

Thus, the resistance was manifested as a reduction in the rate of virus accumulation and, directly or indirectly, virus movement.

When the CMV–resistant cultivar was inoculated with CMV + ZYMV, a synergistic reaction occurred, manifested via both a very severe disease and some breakage of the resistance. Thus, at 5 d.p.i., there was a 4– to 8–fold increase in the level of CMV accumulation in the cotyledons doubly–inoculated with CMV + ZYMV over that which occurred during infection of the CMV–resistant cultivar by CMV alone, and the level of CMV in the doubly–inoculated, first, systemically–infected leaf was 8–fold higher than in the corresponding doubly–infected cotyledon. By comparison, the first leaf of the CMV–resistant plant inoculated by CMV alone did not show detectable accumulation of CMV. By 10 d.p.i., when CMV had accumulated in the first systemically–infected leaf of the CMV–resistant plant, the level of CMV in the plants infected by CMV + ZYMV was 64– to 128–fold higher than in the plants infected by CMV alone at 10 d.p.i. This was still 8–fold less CMV than accumulated in the first systemic leaf of the susceptible cultivar infected by CMV + ZYMV, but was similar to the level of CMV in the first systemic leaf of the susceptible cultivar infected by CMV alone. Thus, the CMV + ZYMV synergy strongly stimulated the levels of accumulation of CMV to the extent that the resistance against CMV was broken. The inability of CMV to accumulate to the same level in the doubly–infected resistant cultivar as in the doubly–infected susceptible cultivar could be a manifestation of some differential effects on CMV cell–to–cell movement in the resistant cultivar, even in the presence of ZYMV.

Discussion

Resistance vs. Synergy vs. Interference

The data shown here demonstrate that some transgenic plants expressing plant viral sequences and exhibiting resistance to the homologous virus can show synergy with unrelated viruses known to interact with the viral progenitor of the transgene. Thus, if these viral transgenes are to be used to protect plants from infection of the corresponding viruses, then a risk–benefit analysis will have to be performed. If the virus that interacts synergistically with the transgene viral progenitor is known not to be present, then the risk could be reduced. However, the use of crop species not susceptible to the synergistic viral partner, or using cultivars of a susceptible species that contain resistance genes may not be sufficient to prevent the synergism, depending on the mechanism of the resistance. The benefits of interference with infection of other unrelated viruses should also be considered in the risk–benefit analysis.

For many viruses there are a number of different viral sequences that can be used to obtain pathogen–derived resistance (Lomonossoff, 1995). Since synergy was observed in only a few cases among the combinations of 11 viral transgenes and four different viruses, and many of those transgenes contained functional rather than defective viral transgenes, it should be possible to select particular viral sequences that provide resistance to the homologous virus, do not exhibit synergy and perhaps also still show some interference with other unrelated viruses.

Most of the examples of interference also involved viral transgenes that encoded functional viral proteins, although two cases (TMV–54K and PVY–NIb genes) involved viral transgenes that showed extreme resistance to the homologous virus. Since these same lines showed interference with viruses known to interact synergistically with TMV and PVY, it would seem likely that transgenic plants expressing both the TMV–54K gene and the PVY–NIb gene would provide resistance to both viruses, interfere with PVX and not show synergy with any of the four viruses tested. This strategy is likely to be successful with pathogen–derived resistance genes that impart extreme resistance to the homologous viral pathogen, but not with such viral transgenes that only delay infection or show a tolerance of infection by the homologous virus. In the latter instances, the resistance might readily be broken by synergy.

Natural Resistance vs. Pathogen–Derived Resistance

The ability of synergy between CMV + ZYMV to break the resistance against infection by CMV showed that the use of natural resistance genes against viral pathogens also has associated risks, if those resistance genes do not exhibit extreme resistance to the corresponding virus. Most natural resistance to virus infection, which has been utilized, involves reductions in virus accumulation and virus–induced disease (tolerance). Therefore, such plants are still capable of interacting synergistically with other viruses (Anderson *et al.*, 1996). Moreover, as shown here, the resistance against one of the viral partners of the synergy can actually be broken when that virus is the one for which accumulation is increased. These data indicate that the use of tolerance genes to virus disease, widely accepted in agriculture, also carry risks associated with their use. In some cases, these risks are much higher than would occur if pathogen-derived resistance was used instead. In the latter case, the risk of synergy can even be eliminated by choosing particular viral segments.

Acknowledgements

This work was supported by BARD grant No. US-2541-95R. P.P. was supported by the Scottish Office of Agriculture, Environment and Fisheries Department.

References

Anderson EJ, Stark DM, Nelson RS, Powell PA, Tumer NE, Beachy RN (1989) Transgenic plants that express the coat protein genes of tobacco mosaic virus or alfalfa mosaic virus interfere with disease development of some nonrelated viruses. Phytopathology 79:1284–1290

Anderson JM, Palukaitis P, Zaitlin M (1992) A defective replicase gene induces resistance to cucumber mosaic virus in transgenic tobacco plants. Proc Natl Acad Sci USA 89:8759–8763

Anderson EJ, Kline AS, Morelock TE, McNew RW (1996) Tolerance to blackeye cowpea mosaic potyvirus not correlated with decreased virus accumulation or protection from cowpea stunt disease. Plant Dis 80:847–852

Audy P, Palukaitis P, Slack SA, Zaitlin M (1994) Replicase-mediated resistance to potato virus Y in transgenic tobacco plants. Mol Plant-Microbe Interact 7:15–22

Cooper B, Lapidot M, Heick JA, Dodds JA, Beachy RN (1995) A defective movement protein of TMV in transgenic plants confers resistance to multiple viruses whereas the functional analog increases susceptibility. Virology 206:307–313

Gal-On A, Kaplan I, Palukaitis P (1995) Differential effects of satellite RNA on the accumulation of cucumber mosaic virus RNAs and their encoded proteins in tobacco vs zucchini squash with two strains of CMV helper virus. Virology 208:58–66

Golemboski DB, Lomonossoff GP, Zaitlin M (1990) Plants transformed with a tobacco mosaic virus nonstructural gene sequence are resistant to the virus. Proc Natl Acad Sci USA 87:6311–6315

Goodman RM, Ross AF (1974) Enhancement of potato virus X synthesis in doubly infected tobacco occurs in doubly infected cells. Virology 58:16–24

Hull R, Davies JW (1992) Approaches to nonconventional control of plant virus diseases. Crit Rev Plant Sci 11:17–33

Kassanis B (1963) Interactions of viruses in plants. Adv Virus Res 10:219–255

Lomonossoff GP (1995) Pathogen-derived resistance to plant viruses. Annu Rev Phytopathol 33:323–343

Poolpol P, Inouye T (1986) Enhancement of cucumber mosaic virus multiplication by zucchini yellow mosaic virus in doubly infected cucumber plants. Ann Phytopath Soc Japan 52:22–30

Scholthof K-BG, Scholthof HB, Jackson AO (1993) Control of plant virus diseases by pathogen-derived resistance in transgenic plants. Plant Physiol 102:7–12

Vance VB, Berger PH, Carrington JC, Hunt AG, Shi XM (1995) 5' Proximal potyviral sequences mediate potato virus X/potyviral synergistic disease in transgenic tobacco. Virology 206:583–590

Wilson TMA (1993) Strategies to protect crop plants against viruses: Pathogen-derived resistant blossoms. Proc Natl Acad Sci USA 90: 3134–3141

Risk Assessment of Transgenic Plants Expressing the Coat Protein Gene of Plum Pox Potyvirus (PPV)

Edgar Maiss[1], Mark Varrelmann[1], Chris DiFonzo[2] and Benjamin Raccah[2]

[1]University of Hannover, Herrenhaeuser Str. 2, 30419 Hannover, Germany

[2]Department of Virology, The Volcani Center, Post Box 6, Bet Dagan 50–250, Israel

Abstract

During the last years numerous plants have been transformed with viral genes, especially coat protein genes (CP), to establish entirely new forms of resistance, the so called pathogen–derived resistance. This new strategy has, however, also raised concerns about the possibility of heterologous encapsidation and recombination in transgenic plants. In the present study we report on the incidence of heterologous encapsidation, recombination and complementation in plants expressing the CP gene of an aphid transmissible isolate (AT) of plum pox potyvirus (PPV). In addition, four different mutations of the CP gene, probably affecting particle assembly, were generated within an infectious full–length clone of PPV to investigate their effect on virus infection and symptom expression on non–transgenic and transgenic plants. In order to create biologically safe transgenic plants, the same mutated genes were used for the transformation of *Nicotiana benthamiana*. The use of CP genes of PPV in transgenic plants led to transcapsidation of several potyviruses. Although, in addition, only homologous recombination and complementation of mutated PPV has been observed in transgenic plants untill now, the consequences of using native coat protein genes of potyviruses for the production of transgenic plants should be evaluated thoroughly.

Introduction

Plum pox potyvirus (PPV), which is widespread all over Europe, is a member of the potyviridae and the causal agent of Sharka disease. PPV particles are flexuous rods 12 nm in diameter and 700–900 nm in length. Within virus particles a single–stranded positive–sense RNA molecule of about 10 kb is found. The RNA has a genome–linked protein (VPg) covalently linked to its 5'–end and a poly(A) tract at the 3'–end. The structure and expression of the genome as well as the processing of the polyprotein, have been described by Riechmann *et al.* (1992).

Infections of plum, peach and apricot with PPV give raise to severe losses in yield and quality of fruits. Recently a new isolate of PPV (PPV–SoC), which infects systemically sour and also sweet cherry, has been reported from Moldova and Bulgaria. PPV is transmitted in a non–persistent manner by different aphids. Due to its transmission mode, the lack of suitable resistance genes and time consuming breeding procedures, efficient control of PPV is very difficult to achieve. Therefore new strategies of pathogen derived resistance have been developed for protection against PPV.

During the last decade numerous plants have been transformed with viral genes, especially coat protein (CP) genes, for the generation of resistance. For establishment of coat protein mediated resistance against PPV, the complete coat protein gene of PPV under control of the CaMV 35S promoter and the polyadenylation signal of the nopaline synthetase gene was inserted into the genome of *Nicotiana benthamiana* via Agrobacterium-mediated transformation (Timpe *et al.* 1992). Transformants were regenerated and self pollinated for the production of homozygous plant lines. One line (27–4) was used for resistance tests to PPV. After inoculation with PPV, plants of line 27–4 developed systemic symptoms, but later on recovered from infection, since new developing leafs were free of symptoms. In addition, no virus could be detected by means of ELISA from symptomless leaves. Despite the fact that the new strategy promises to be of great agronomic and economic use, it has, however, also raised concerns about potential biological and environmental risks.

Before large scale use of transgenic, virus-resistant plants containing viral genes, the possible impact of molecular interactions with challenging viruses leading to heterologous encapsidation, complementation and/or recombination has to be considered.

Heterologous encapsidation between two related viruses occurs in nature in mixed infections (Wen and Lister, 1991). In transgenic plants, expressing functional viral coat proteins in every cell, these CPs can be used by closely related incoming viruses for heterologous encapsidation as shown by Farinelli *et al.* (1992) and Maiss *et al.* (1994) for different potyviruses. With this transgene coat protein functions like host- or vector-specificity can be transferred to the invading virus (Lecoq *et al.* 1993; Lecoq *et al.* 1994).

Homologous and heterologous recombination between viral RNAs (Bujarski and Kaesberg, 1986) or between viral and host RNA (Mayo and Jolly, 1991) is thought to be an important source of variability in viral RNAs, and may have played an important role in virus evolution. In transgenic plants containing a functional viral RNA in every cell, the probability for a recombination event between transgenic RNA and incoming viral RNA is considered to be higher than in non-transgenic plants or in mixed infections. Greene and Allison (1994) have found evidence for recombination between transgenic viral plant transcripts and defective invading viral RNA. Recombination between a defective CaMV genome and transgenic host plants containing the missing viral gene was demonstrated by Gal *et al.* (1992) and Schoelz and Wintermantel (1993).

On the other hand, defective genes of a virus mutant challenging a transgenic plant can be complemented by the functional viral protein expressed in this plant, as demonstrated by Holt and Beachy (1991) for the tobamoviruses and by Schoelz and Wintermantel (1993) for the caulimoviruses.

Results

In this study the aphid transmission of normally non-aphid-transmissible (NAT) isolates of zucchini yellow mosaic potyvirus (ZYMV) and tobacco vein mottling potyvirus (TVMV) from transgenic plants expressing the coat protein from the aphid transmissible isolate of PPV (PPV-AT) was investigated. In the same way as demonstrated for the tobamo- and caulimoviruses, the occurrence of complementation of coat protein defective PPV-NAT in transgenic plants expressing the functional coat protein of the PPV-AT isolate has been observed. In addition, it was found that homologous recombination events between transcripts of a viral transgene and defective invading viral RNA of PPV and a PPV-ZYMV chimera occur in transgenic plants.

Heterologous encapsidation and aphid transmission

N. benthamiana plants of line 27-4 (Timpe *et al.*, 1992) expressing the CP of PPV-AT were checked for their ability for heterologous encapsidation with different viruses and for support of aphid transmissibility of two non-aphid transmissible viruses (ZYMV-NAT and TVMV-NAT). Whereas high frequencies of heterologous encapsidation have been observed with potyviruses, no heterologous encapsidation occurred with potato X potexvirus (PVX), tobacco mosaic tobamovirus (TMV) and beet necrotic yellow vein virus (BNYVV; Maiss *et al.*, 1995). To test the aphid transmission efficiency of heterologous encapsidated virus particles from *N. benthamiana* plants, the first step was aimed at attaining a consistent mechanical infection of ZYMV-NAT in a non-transgenic line (27-1) and in the transgenic line (27-4) with ZYMV-NAT (Table 1). Surprisingly, great difficulty was encountered in attempting to inoculate *N. benthamiana* with the virus. In all cases, no systemic spread was recorded from the inoculate leaves challenged with ZYMV-infected squash extracts. Partially successful systemic infection was only obtained when locally infected *N. benthamiana* leaves served as the virus source. Plants testing positive for systemic infection of ZYMV-NAT were used as a source for aphid transmission. Green peach aphid, *Myzus persicae* (Sulzer) has been used for transmission. 20 aphids were used on each of 164 test plants. No ZYMV-NAT transmission occurred from control plants (27-1), nor were transmissions recorded from *N. benthamiana* plants expressing the PPV-AT CP (27-4). Experiments were continued using a non-aphid transmissible strain of TVMV-NAT, (provided by J. Shaw, University of Kentucky, USA). Only a single infection from one 27-4 plant was recorded out of the 44 test plants exposed to 30 aphids on each.

Complementation and recombination

The core region of the CP gene of PPV-NAT, like that of other potyviruses, contains two amino acid motifs supposed to be responsible for particle assembly (Jagadish *et al.* 1993; Dolja *et al.* 1994).

Table 1. Aphid transmission of normally non-aphid-transmissible (NAT) isolates of ZYMV and TVMV from non-transgenic (27-1) and transgenic plants (27-4)

Virus isolate	Transgenic source plant	n aphids per test plant	n test plants	n infected
ZYMV-NAT	27-1	20	21	0
	27-4	20	164	0
ZYMV control		5	50	47 (94%)
TVMV-NAT	27-1	40	5	0
	27-4	30	44	1 (2%)
PPV-NAT	27-1	10	20	0
	27-4	10	20	0

These motifs were rendered nonfunctional by replacing the original amino acids with ones of different chemical properties (Fig. 1). These two mutations were introduced separately and together into the CP gene. The three mutated CP genes were introduced into the full-length clone of PPV-NAT (Maiss *et al.* 1992) and tested for infectivity on transgenic (27-4) and non-transgenic (27-1) *N. benthamiana* plants via particle bombardment (Gray *et al.* 1994; Gal-On *et al.* 1995).

None of these mutants produced systemic infections in non-transgenic *N. benthamiana* plants, but two of them infected 3 out of 10 (3/10) and 9/10, respectively, transgenic *N. benthamiana* plants expressing the CP-AT of PPV. It was not possible to transmit these mutants to non-transgenic *N. benthamiana,* but transmission was possible to transgenic *N. benthamiana* (2/3 and 6/9 plants, respectively) in the first and second passage (Table 2). Only when transmitted in the third passage to non-transgenic *N. benthamiana*, one out of six plants inoculated with the RQ>DV mutant displayed systemic infections. The CP gene was amplified with specific primers via IC-RT-PCR (Wetzel *et al.* 1992) and cloned into appropriate vectors. Several clones from these PCR-products were analyzed with appropriate restriction enzymes for the presence of the introduced mutation (DV). 7/27 analyzed clones still contained the DV mutations, the others did not.

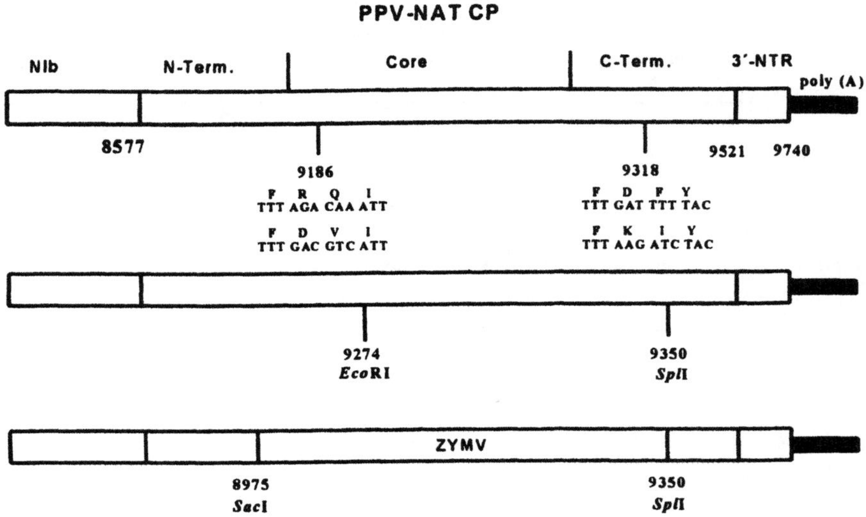

Fig. 1. Different mutations in the CP gene of PPV–NAT (nucleotide positions according to Maiss *et al.*, 1992)

These results showed that the transgenic CP–AT complemented the defective CP–NAT in both cases, even after 2 or 3 passages, respectively. In addition, a recombination event has occurred between the RNA of the incoming virus and the transgenic RNA in one of six plants in the case of the RQ>DV CP–mutant.

Furthermore, two frame–shift mutants were generated within the PPV CP gene at an *Eco*RI and *Spl*I restriction site, respectively, by a fill–in reaction with Klenow–Polymerase (Boehringer) and one deletion–mutant by removing the region between *Eco*RI and *Spl*I without destroying the open reading frame (Fig. 1). These mutants were also tested for their ability to infect non–transgenic and transgenic *N. benthamiana* plants (Table 2). None of them was able to give a systemic infection in non–transgenic plants. In contrast, the *Spl*I frame–shift mutant was able to produce a systemic infection in 4/10 transgenic *N. benthamiana*. Virus particles from one plant displayed the ability to infect subsequently non–transgenic *N. benthamiana*. In a repetition of this assay, 3/10 transgenic *N. benthamiana* became systemically infected and recombinant virus was transmitted subsequently to transgenic and non–transgenic *N. benthamiana* with 100% infection rate. The recombination event was confirmed by sequencing, showing that the original PPV sequence was exactly restored.

In a third approach, the core region of the CP gene of PPV–NAT was replaced by the equivalent part of the core region of ZYMV, between the restriction sites *Sac*I and *Spl*I (Fig. 1), resulting in a chimeric CP with a functional open reading frame and both amino acid motifs responsible for particle assembly from ZYMV.

Table 2. Infectivity of different PPV-CP mutants on transgenic (27-4) and non-transgenic (27-1) *N. benthamiana*. Appearance of viable virus from PPV-CP mutants after passage from transgenic to non-transgenic *N. benthamiana*. Passages were done 21 days post inoculation.

	particle bombardment		1. passage[3]		2. passage[4]		3. passage[5]	
mutant/plant	27-4	27-1	27-4	27-1	27-4	27-1	27-4	27-1
RQ>DV	9[1]/10[2]	0/20	6/9	0/9	6/6	0/6	5/6	1/6
DF>KI	0/10	0/19						
RQ>DV+DF>KI	3/10	0/10	2/3	0/3	2/2	0/2	2/2	0/2
*Eco*RI/*Spl*I deletion	0/10	0/10						
*Eco*RI frame-shift	0/10	0/20						
*Spl*I frame shift	4/10 3/10	0/10 0/10	3/4 3/3	3/4 3/3				
ZYMV-CP chimera	10/10	0/10	10/10	0/10	10/10	10/10		

1 plants infected

2 plants inoculated

3 sap of single 27-4 plants from the particle bombarded plants was inoculated to transgenic (27-4) and non-transgenic (27-1) *N. benthamiana* plants

4 sap of single 27-4 plants from the 1. passage was inoculated to transgenic (27-4) and non-transgenic (27-1) *N. benthamiana* plants

5 sap of single 27-4 plants from the 2. passage was inoculated to transgenic (27-4) and non-transgenic (27-1) *N. benthamiana* plants

The chimera was also tested for its ability to infect non-transgenic and transgenic plants. Infectivity assays exhibited no systemic infections in non-transgenic *N. benthamiana*, but the chimeric virus was able to infect transgenic *N. benthamiana* plants to 100% (Table 2). It was not possible to transmit putative "recombinant virus" to non-transgenic *N. benthamiana* plants within the first passage, but this was possible on transgenic plants in all cases. Therefore, the CP of the transgenic plant complemented the defective one. The second passage led to systemically infected non-transgenic *N. benthamiana* plants, indicating that in addition to complementation a recombination event has taken place. After IC-RT-PCR several CP clones from the infected non-transgenic plants were analysed for the presence of the inserted ZYMV CP core. As shown for the CP assembly mutant

(RQ>DV) a portion of the analysed clones still contained the replaced part of ZYMV.

Discussion and conclusions

Aphid transmission of a non–aphid–transmissible strain of TVMV, which has been heterologous encapsidated in the coat protein of an aphid transmissible strain of PPV, has been observed from transgenic plants. This result is in accordance with findings of Lecoq *et al.* (1993), who have reported the transmission of a ZYMV-NAT strain from transgenic plants expressing also a CP of a PPV–AT isolate. Nevertheless, the transmission rate of TVMV–NAT from transgenic plants was lower than with ZYMV–NAT, as observed by Lecoq *et al.* (1993). Obviously, the coat protein and the functions attaching thereto can be supplemented to a certain degree between nearly related viruses. In contrast to Lecoq *et al.* (1993), we were not able to confirm the aphid transmission of ZYMV–NAT from transgenic *N. benthamiana* 27–4, although the CP was clearly detected. The reasons that may account for the failure to attain aphid transmission in our laboratory, could be a certain "filtering" of ZYMV lines that occurs in the process of the virus adjustment to *N. benthamiana.* Another more remote possibility could be the difference in the individual ability in virus transmission of the *Myzus persicae* clones in the different laboratories. Heterologous encapsidation could stimulate virus spread when transgenic crops expressing aphid transmissible coat proteins will be grown in large areas. However, the frequency of aphid transmission of heterologous encapsidated particles is much lower than the transmission of wild–type virus. The putative risks of heterologous encapsidation and aphid transmission can be minimized or completely circumvented in different ways. One possibility is the use of transgenic plants that are transformed with non–structural genes, or by using CP genes that do not support particle assembly or aphid transmission. For this purpose, modified CP genes of PPV (see Fig. 1), which have been used in the context of the PPV–NAT full–length clone, have also been transformed into *N. benthamiana* lines, to test their ability to confer resistance without heterologous encapsidation or aphid transmission.

Homologous recombination events between the transcripts of a virus–resistant transgenic plant and defective invading viral RNA have been detected at a high frequency, after a period of complementation, using several CP–defective PPV-NAT mutants in transgenic plants expressing the functional coat protein of PPV-AT. However, it should be taken into account that in all experiments described, recovery resistant plants were used. These plants allow an incoming virus to replicate before the resistance mechanism affects virus movement and/or replication. For experiments to detect recombination, CP–defective PPV mutants were used. The mutants have a high selection pressure for recombination, and it can be assumed that such mutants will normally not occur in the field. Therefore, it is difficult to predict whether the observed recombination events will also take place within field release experiments of plants expressing functional potyvirus coat proteins. Nevertheless, functional viral genes in transgenic plants may have several

disadvantages, which could be minimized by utilising only defective or non-translatable viral genes for the generation of pathogen-derived resistance. The usefulness of this concept for pathogen-derived resistance is under investigation.

References

Bujarski JJ, Kaesberg P (1986) Genetic recombination between RNA components of a multipartite plant virus. Nature 321:528–531

Dolja VV, Haldeman R, Robertson NL, Dougherty WG, Carrington JC (1994) Distinct functions of capsid protein in assembly and movement of tobacco etch potyvirus in plants. EMBO J 13:1482–1491

Farinelli L, Malnoe P, Collet GF (1992) Heterologous encapsidation of potato virus Y strain 0 (PVY^{0}) with the transgenic coat protein of PVY strain N (PVY^{N}) in Solanum tuberosum cv. Bintje. Bio/Technology 10:1020–1025

Gal S, Pisan B, Hohn T, Grimsley N, Hohn B (1992) Agroinfection of transgenic plants leads to viable cauliflower mosaic virus by intermolecular recombination. Virology 187:525–533

Gal-On AG, Meiri E, Huet H, Hua WJ, Raccah B, Gaba V (1995) Particle bombardment drastically increases the infectivity of cloned DNA of zucchini yellow mosaic potyvirus. J Gen Virol 76:3223–3227

Gray DJ, Hiebert E, Lin CM, Compton ME, McColley DW, Harrison RJ, Gaba VP (1994) Simplified construction and performance of a device for particle bombardment. Plant Cell Tissue Organ Cult 37:179–184

Greene AE, Allison RF (1994) Recombination between viral RNA and transgenic plant transcripts. Science 263, 1423–1425

Holt CA, nd Beachy RN (1991) In vivo complementation of infectious transcripts from mutant tobacco mosaic virus cDNAs in transgenic plants. Virology 181:109–117

Jagadish MN, Huang DX, Ward CW (1993) Site-directed mutagenesis of a potyvirus coat protein and its assembly in Escherichia coli. J Gen Virol 7: 893–896

Lecoq H, Ravelonandro M, Wipf-Scheibel C, Monsion M, Raccah B, Dunez J (1994) Significance of the heterologous encapsidation of zucchini yellow mosaic potyvirus in transgenic plants expressing plum pox potyvirus capsid protein. Bulletin OEPP 24:555–559

Lecoq H, Ravelonandro M, Wipf-Scheibel C, Monsion M, Raccah B, Dunez J (1993) Aphid transmission of a non-aphid-transmissible strain of zucchini yellow mosaic potyvirus from transgenic plants expressing the capsid protein of plum pox potyvirus. Mol Plant-Microbe Interact 6:403–406

Maiss E, Timpe U, Brisske A, Lesemann DE, Casper R (1992) Infectious in vivo transcripts of plum pox potyvirus full-length cDNA clone containing the cauliflower mosaic virus 35S RNA promoter. J Gen Virol 73:709–713

Maiss E, Koenig R, Lesemann D-E (1995) Heterologous encapsidation of viruses in transgenic plants and in mixed infections. 3rd International Symposium on The biosafety results of field tests of genetically modified plants and microorganisms, Monterey, California United States

Mayo M, Jolly C (1991) The 5' terminal sequence of potato leaf roll virus RNA: Evidence of recombination between virus and host RNA. J Gen Virol 72:2591–2595

Ravelonandro M, Varveri C, Delbos R, Dunez J (1988) Nucleotide sequence of the capsid protein gene of plum pox potyvirus. J Gen Virol 69:1509–1516

Riechmann JL, Lain S, Garcia JA (1992) Highlights and prospects of potyvirus molecular biology. J Gen Virol 73:1–16

Schoelz JE, Wintermantel WM (1993) Expansion of viral host range through complementation and recombination in transgenic plants. Plant Cell 5:1669–1679

Timpe U, Maiss E, Landsmann J, Casper R (1992) Coat protein mediated cross protection" gegen das Scharka Virus. Mitteilungen der Biologischen Bundesanstalt für Land- und Forstwirtschaft 238, 212.

Wen F, Lister RM (1991) Heterologous encapsidation in mixed infections among four isolates of barley yellow dwarf virus. J Gen Virol 72:2217–2223

Wetzel T, Candresse T, Macquaire G, Ravelonandro M, and Dunez J (1992) A highly sensitive immunocapture polymerase chain reaction method for plum pox potyvirus detection. J Virol Methods 39:27–37

Mixed Infections and Genetic Exchange Occur in Natural Populations of Cucumber Mosaic Cucumovirus

Fernando García-Arenal[1], José Luis Alonso-Prados[1], Miguel A. Aranda[1], José M. Malpica[2], and Aurora Fraile[1]

[1] Depto. de Biotecnología, E.T.S.I. Agrónomos, Universidad Politécnica de Madrid, 28040 Madrid, Spain.

[2] Area de Biología Molecular y Virología, Centro de Investigación y Desarrollo, Instituto Nacional de Investigaciones Agrarias, 28040 Madrid, Spain.

Abstract

A potential risk of transgenic plants expressing viral sequences derives from the possibility of new viral genotypes being generated by recombination between the genomic RNA of infecting viruses and RNA transcribed from the transgene. It is important to know if the probability of generation of new variants through this mechanism is higher than through recombination between natural variants in mixed infections. To date, no quantitative data have been reported on the frequency of mixed infections or on the frequency of recombination in natural populations of a plant virus. We will report such an estimation derived from the analysis of about 250 field isolates of cucumber mosaic cucumovirus (CMV), representing different epidemic episodes on different host plants in Spain. Isolates were characterized by the pattern of fragments in ribonuclease protection assays of probes representing different regions of the three genomic RNAs. For each probe, only three well defined patterns (haplotypes) were found. this analysis permitted estimation of the frequency of host plants with mixed infections of different haplotypes to be at least 15%. From these mixed infections new types could arise by genetic enchange, either by recombinantion or by reassortment. The frequency of genetic exchange in the analyzed CMV populations has also been estimated.

Introduction

Transgenic plants that express a segment of a plant virus genome are often resistant to that specific virus. Although different viral genes have been shown to confer transgenic protection against virus infection for several virus systems, the replicase and coat protein genes are those that appear to be more useful (Lomonossoff, 1995). Although the use of this type of resistance has enormous potential benefits, its use is being delayed because of concerns about potential hazards to agriculture and to the environment. The main potential hazard is considered to be the possibility for producing new viruses as a result of the widespread use of these plants. These new viruses would be generated by recombination between the genome of a virus infecting the transgenic plant and transcripts from the virus-derived transgene (Allison *et al.* 1996).

Recombination has been shown to play an important role on the evolution of viruses having both DNA and RNA (the vast majority of those infecting plants) genomes (Lai, 1996). Recombination has also been shown to occur in transgenic plants between challenge virus and transgenes under experimental situations, and the new generated virus often showed novel pathogenic traits (Allison *et al.*, 1996). Nevertheless, there is a remarkable scarcity of validated data on which to estimate the risks of appearance of new viruses because of the extended use of transgenic plants.

Data from which to make risk assessments must necessarily include studies of natural virus populations. It is well known that the selective forces driving virus evolution may be very different under natural (= field) conditions, as compared to what can be observed in the laboratory or greenhouse. Also, it is important to know whether the risk of generation of new viruses by recombination between challenging viruses and transgenes is larger that the probability of generation of new viruses by recombination between two different viruses co-infecting a non-transgenic plant. We present here data on the frequency of mixed infections, and on the frequency of genetic exchange, in natural populations of the plant RNA virus, cucumber mosaic cucumovirus (CMV). To our knowledge these are the first quantitative data on these issues reported for any plant virus. Part of the data we present here have been published recently (Fraile *et al.*, 1997).

The system analyzed

CMV is a plant virus with a tripartite, messenger sense, single-stranded RNA genome. Replication-related functions are encoded by RNAs 1 and 2; movement related functions are encoded by RNAs 2 and 3; the coat protein (CP), which also determines virus transmission by its aphid vectors, is encoded by RNA3. CMV has a very wide host range, is efficiently transmitted by numerous species of aphids, and is of paramount economic importance in many crops. CMV isolates have been classified into two subgroups according to the degree of similarity of their genomic RNAs (for a review on CMV see Palukaitis *et al.*, 1992). Because of the tripartite nature of its genome, genetic exchange between CMV strains might occur by RNA recombination and by reassortment of the genomic segments. Indeed it has been proposed that multipartitism has evolved in RNA viruses to favour genetic exchange (Chao, 1988).

CMV is endemic in Spain, where it induces frequent epidemics in certain crops (Luis-Arteaga *et al.*, 1988). We have charactierized 279 CMV isolates sampled between 1989 and 1995 at different locations representing important agricultural areas in Northern, Eastern and Central Spain. Isolates were obtained from the major affected crops (Table 1). Virions were directly purified from field-infected material, or after a single passage in tobacco in the greenhouse. From these isolates virion RNA was typified by the pattern of fragments obtained in ribonuclease protection assays (RPA) of RNA probes complementary to regions (500-1000 nucleotides long) in RNA 1, in RNA 2 and in both open reading frames (for the movement protein and the CP) in RNA 3 (for details see Fraile *et al.,* 1997). The use of probes derived from isolates Fny-CMV (in subgroup I of CMV strains) or Ls-CMV (in subgroup II) allowed the unequivocal identification of each field isolate as belonging to subgroups I or II.

Table 1. Sampled natural populations of CMV

PLACE	YEAR	HOST PLANT	No. ISOLATES
Valencia	1989	Tomato	17
	1990	Tomato	19
	1991	Tomato	11
	1994	Tomato	7
	1995	Tomato+Melon	14
Barcelona	1992	Tomato	36
	1993	Tomato+Weeds	39
	1994	Tomato	16
	1995	Melon	5
Murcia	1995	Melon	6
Zaragoza	1994	Melon	10
	1995	Melon	12
Madrid	1993	Tomato+Cucurbits	25
	1994	Pepper+Cucurbits+Bean	16
	1995	Melon	8
Badajoz	1995	Melon	14
León	1993	Pepper	24
TOTAL			279

For all of subgroup I isolates RPA analyses yielded two main patterns, for each of the four cRNA probes: pattern A in which large fragments similar in size to the probe were protected, and pattern B in which only smaller fragments were protected. No intermediate patterns were found for any of the analyzed isolates. Mixes of patterns A and B, representing double infections of the original field source plant, were easy to identify (not shown, but see Fraile *et al.,* 1997).

Genetic structure of CMV populations: mixed infections and genetic exchange

Subgroup II of CMV represented 12% of the analyzed isolates, while 88% belonged to subgroup I. RPA pattern A or B, or a mixed pattern AB, for each of the four probes, was used to define haplotypes within subgroup I isolates; 20 different haplotypes were so defined (Table 2). 71% of subgroup I isolates showed either an A or B pattern for all four probes, the frequency of haplotypes 1 (A, A, A–A) and 2 (B, B, B–B) was almost the same.

Table 2. Frequencies for genetic types of CMV in field populations

	RPA pattern				Frequency	
Type[a]	RNA 1	RNA 2	RNA 3	Number[b]	whole sample	Sg. I
Sg.II				32	0.12	
Sg.I					0.88	
1	A	A	A–A	90		0.36
2	B	B	B–B	86		0.35
3	**AB**	**AB**	**AB–AB**	**6**		
4	**A**	**AB**	**AB–AB**	**2**		
5	**B**	**AB**	**AB–AB**	**1**		
6	**AB**	**A**	**AB–AB**	**8**		
7	**AB**	**A**	**A–A**	**4**		
8	**A**	**AB**	**A–A**	**13**		
9	**AB**	**A**	**B–B**	**1**		
10	**AB**	**B**	**B–B**	**5**		
11	**B**	**AB**	**B–B**	**1(41)**		**0.17**
12	*A*	*B*	*A–A*	*3*		
13	*A*	*B*	*B–B*	*4*		
14	*A*	*A*	*B–B*	*3*		
15	*B*	*A*	*B–B*	*4*		
16	*B*	*A*	*A–A*	*2(16)*		*0.06*
17	***A***	***A***	***B–A***	***8***		
18	***A***	***A***	***A–B***	***1***		
19	***B***	***B***	***B–A***	***4***		
20	***B***	***B***	***A–B***	***1(14)***		***0.06***
TOTAL				279		

[a] Within subgroup I, homogeneous types (1–2) are shown in normal type, types due to mixed infection (3–11) in bold type, reassorted types (12–16) in italics, and recombinant types (17–20) in bold italics.
[b] Numbers in parentheses are subtotals of the different groups of types.

Types 3 to 11 represent mixed infections with isolates of types 1 and 2, in which both types coexist for the three genomic RNAs, or either pattern A or pattern B has outcompeted the other for RNA 1, 2 or 3. Mixed infections represent 17% of the analyzed subgroup I isolates. Because of the method of analysis, this estimate is a lower limit to the frequency of mixed infections in the field.

Types 12 to 16 represent reassortant genomes generated by the displacement of RNAs of pattern A or B by the other, for any of RNAs 1, 2 or 3, from originally mixed infections. Reassortant genomes occur in 6% of subgroup I isolates. It is important to note that only 7 of the 8 possible combinations of the genomic segments (types 1, 2, 11–16) were found, and at very different frequencies. Also, each reassortant type was found only in one population (not shown). Reassortants between subgroups I and II were not found. These results indicate that reassortants did not compete efficiently with the parental types.

Because two probes were used for RNA 3, recombinants at this RNA could be identified. Recombination between the 5' and the 3' halves of RNA3 occurred in 6% of subgroup I isolates. Again, the frequencies of the two possible recombinants (A–B or B–A) were very different, and recombinant genomes did not become prevalent in the population (not shown).

Thus, our results show that in spite of the frequent occurrence of mixed infections in subgroup I isolates, of haplotypes 1 and 2 (≥ 17%) and, regardless of the (undetermined) frequency of genetic exchange, isolates with reassortant or recombinant genomes are not favoured in natural populations of CMV. Indeed, selection forces, of as yet unknown nature (Fraile *et al.*, 1997) seem to favour the homologous association of types A, or B, for all three genomic RNAs. Interestingly, a much higher frequency of recombinants was found for CMV satellite RNAs in the same analyzed populations (Aranda *et al.*, 1997).

Conclusions

Because of the economic importance of CMV, and because no conventional resistance to it has been described in most crops, the use of resistant transgenic plants that express CMV sequences would be very useful. The results we present here show that, for CMV, new genetic types generated by genetic exchange are at competitive disadvantage in natural populations. Thus, it could be concluded that the use of transgenic plants expressing CMV sequences would have low ecological risks. This conclusion must be taken with caution, since the fitness of a genotype depends on the environment, and varies with environmental changes.

Acknowledgements

This work was in part supported by grant PB93–038, MEC, Spain.

References

Allison RF, Schneider WL, Greene AE (1996) Recombination in plants expressing viral transgenes. Sem Virol 7:417–422

Aranda MA, Fraile A, Dopazo J, Malpica JM, García-Arenal F (1997) Contribution of mutation and RNA recombination to the evolution of of a plant pathogenic RNA. J Mol Evol 44:81–88

Chao L (1988). Evolution of sex in RNA viruses. J Theor Biol 133:99–112

Fraile A, Alonso-Prados JL, Aranda MA, Bernal JJ, Malpica JM, García-Arenal F (1997) Genetic exchange by recombination or reassortment is infrequent in natural populations of a tripartite RNA plant virus. J Virol 71:934–940

Lomonossoff GP (1995) Pathogen-derived resistance to plant viruses. Annu Rev Phytopathol 33:323–343

Luis-Arteaga M, Rodríguez-Cerezo E, Maestro C, García-Arenal F (1988) Detection and characterization of an isolate of cucumber mosaic virus infecting borage (*Borago officinalis* L.) in Spain. Plant Dis 72:265–267

Palukaitis, P, Roossinck MJ, Dietzgen RG, Francki RIB (1992) Cucumber mosaic virus. Adv Virus Res 41:281–348

The Potential of a Beneficial Satellite RNA of Cucumber Mosaic Cucumovirus to Acquire Deleterious Functions : Nature *Versus* Greenhouses

Donato Gallitelli, Francesco Grieco and Fabrizio Cillo
Dipartimento di Protezione delle Piante dalle Malattie, Università degli Studi and Centro di Studio del CNR sui Virus e le Virosi delle Colture Mediterranee, Bari, Italy.

Abstract

Part of the notoriety of cucumber mosaic cucumovirus (CMV) lies in its ability to support replication of a small (332–405 nucleotide) satellite RNA (satRNA) which can sometimes affect CMV replication, pathogenesis and symptom expression. Variants with attenuating effects (benign satRNAs) have been extensively studied as potential or actual biocontrol agents for the prevention of CMV-induced diseases. This approach has become the basis for experimental and practical field tests where satRNA-mediated protection has been achieved successfully by either preinoculation of seedlings (cross-protection or "vaccination") or by transgenosis, and apparently no adverse effects have been detected.

The recurrent epidemics of CMV in Italy have offered the opportunity to carry out phylogenetic analysis of several wild-type variants of CMV satRNA in a "natural laboratory". The population of CMV satRNA that we have studied since 1987 has had a propensity to maintain its original genetic heterogeneity from one year to the other. It seems that: i) molecules of the same lineage have limited possibilities to accumulate substitutions; ii) regions of the molecules with the possibility of accumulating substitutions are very likely the same in all the lineages; iii) the data are in agreement with the proposed maintenance of a functional molecular structure as a constraint on CMV satRNA evolution.

Although in principle we agree with the concept that benign CMV-SatRNAs should be used as biological control agent with extreme care, it should also be taken into account that: i) risks pointed out for a given satRNA variant may not be applicable to others; ii) on the basis of the long-term tests carried out in China, data obtained from nature do not seem to fit completely greenhouse experiments. In this view, a satRNA-mediated cross-protection strategy with a pseudorecombinant strain of CMV and several constructs to be used for transformation of tomato plants are under development in our laboratory. The results of some field and laboratory experiments will be briefly reported.

Introduction

Cucumber mosaic virus (CMV), the type species of the genus *Cucumovirus* in the family *Bromoviridae* (Rybicki, 1995), is recognised as a threatening, economically important plant pathogen occurring worldwide. Recently, CMV was associated with tomato diseases of epidemic scale in several countries of the Mediterranean basin (for review, see Gallitelli *et al.* 1995). As widely known, CMV supports the replication of a small (332–405 nucleotide) satellite RNA (satRNA), which can sometimes affect CMV replication, pathogenesis and symptom expression (Collmer and Howell, 1992; Roossink *et al.*, 1992). A number of variants of CMV satRNA intensify symptoms induced by the virus in susceptible plant species, whereas other variants attenuate the symptoms. Variants with attenuating effects (benign satRNAs) have been studied as potential or actual biocontrol agents for the prevention of CMV-induced diseases (see for review Tien and Wu, 1991; Kaper, 1993). This approach was very promising and has become the basis for experimental and practical field tests where satRNA-mediated protection was successfully achieved by either preinoculation of seedlings (cross-protection or "vaccination") or transgenosis (Tien and Wu, 1991; Yie and Tien, 1993). The opposite was demonstrated in other instances, since experimental laboratory evidence was obtained that detrimental satellite molecules can arise from benign molecules by single point mutation (Palukaitis and Roossinck, 1996). This sounds as a clear warning for the safety of CMV satRNA releases in large-scale agriculture. However, it is unknown if the mutation is likely to occur under field conditions and to affect the viral population and, even so, it cannot be predicted if deleterious mutants arising from the benign ones will meet ecological success.

The recurrent CMV epidemics in Italy were the dramatic consequence of the detrimental functions of certain CMV-satRNA variants, but, at the same time, they offered the opportunity to follow the life cycle of satRNAs in nature and to study their dynamics on the basis of wild-type sequence data.

Surveying CMV satRNAs in Italy to understand their life cycle

Although several reports from our country have recorded the occurrence of CMV in different plant species, a really comparable set of data was probably obtained only from surveys carried out from 1988 onwards because: i) the evaluation and collection of samples was carried out regularly and by the same research group; ii) the environmental conditions and agricultural practices were different from one place to another but were basically the same from one year to the other in the same place; iii) biological characterization and nucleic acid technology (molecular hybridization, RT-PCR, sequencing) were applied throughout and in the same laboratory.

Infections supported by individual CMV isolates were most prevalent, both in weeds and cultivated crops. Only in five out of the *c.* 2600 samples analysed was a mixed infection of isolates belonging to different CMV subgroups detected. Obviously, the possibility exists that mixed infections of CMV isolates of the same subgroup escaped detection. Certain weed species (mainly in *Compositae*) seem to play a key role in supplying the primary inoculum for epidemics of the subgroup II CMV isolates that predominate in wild plants throughout winter, and are the most frequently associated with lethal necrosis disease in tomato. The absence of subgroup II isolates in summer could be explained by the natural disappearance of these weeds and by the self-destructive nature of the disease in tomato. As infected tomato plants die, there is a strong reduction of the inoculum available in the environment, and only certain weeds are able to maintain the virus (probably in their seeds) until next autumn.

Subgroup I CMV isolates were more frequently associated with other disease phenotypes in tomato (filimorphism, top stunting, fruit necrosis) and only occasionally with lethal necrosis. On the other hand, they were by far the most common in eggplant, melon, pepper, tobacco, watermelon, zucchini squash and celery, the host in which they overwinter. This suggests that crops rather than weeds may act as preferential reservoirs for Subgroup I isolates.

SatRNA sequences seem to be quite rare in weeds. In Italy, for example, the number of weed species recognized as natural hosts of CMV satRNA in Italy does not exceed 15 (Crescenzi *et al.*, 1993b), suggestings that cultivated species are the major natural reservoirs of satRNA.

These findings are in agreement with results of a phylogenetic analysis of a number of CMV satRNA variants that have been isolated, biologically characterized and completely sequenced in our laboratory. In fact, the original sequence data from wild-type satRNA variants show that they have the propensity to maintain their original heterogeneity from one year to the other (Grieco *et al.*, 1997), and over longer periods when epidemics do not occur, as exemplified by Tfn-satRNA. This molecule was first identified in Campania (southern Italy) in 1989 and was isolated again in Apulia (i.e. more than 300 km away) in 1993 from a commercial tomato field with fruit necrosis symptoms. After five years, Tfn-satRNA retained the original nucleotide sequence with only two substitution (T-172 was changed to C, and G-269 was changed to A).

satRNA-mediated cross-protection

Cross-protection is a mechanism whereby infection of a mild strain of a virus (protecting strain) prevents the subsequent infection or the expression of infection by a virulent strain (challenging strain). With CMV, cross-protection (often referred to as preimmunization or vaccination) is greatly enhanced if the protective strain supports a benign satRNA, because of its ability to prevent infection by CMV strains bearing necrogenic satRNA variants (Kaper, 1995). Although cross-protection represents the most rapid disease control response in cases of sudden CMV outbreak, it raises the problem of potential risks connected with the release of benign satRNA variants in large scale agriculture (Tepfer, 1993; Tepfer and Jacquemond, 1997). Two main problems can be envisaged: i) the transmission by aphids of the protective strain and its satRNA to other susceptible crops where they could cause disease, or to weeds where the satRNA could undergo mutation, and ii) the potential reassortment or true recombination of genome segments (and of satRNA variants) between protecting and challenging virus strains. Although the first risk could be ruled out only if the protective strain is not aphid-transmissible, the second point emerges only if the challenging strain and its satRNA will replicate in cross-protected plants. Different opinions exist about the latter question, which has been repeatedly discussed, although a consensus has not yet emerged (see for review Wilson, 1993).

In our laboratory, a CMV pseudorecombinant was assembled, characterised and used as a biocontrol agent. The helper virus was obtained *in planta* by inoculation of tomato seedlings with CMV–S (a subgroup II strain) genomic RNAs 1+2, CMV–Tfn (a subgroup I Italian isolate) genomic RNA–3, and a biologically active transcript of a cDNA clone of Tfn–satRNA (a 390 nt satRNA supported by CMV–Tfn). This pseudorecombinant, named CMV–ST, was produced because naturally attenuated strains of CMV have not yet been found in Italy, and because it can be monitored with specific riboprobes.

In greenhouse experiments, CMV–ST protected tomato plants against CMV–PG, a subgroup II Italian isolate supporting a necrogenic satRNA, and against other virulent strains of CMV and their satRNAs. Experimentally, CMV–ST did not induce detectable symptoms in zucchini squash, cucumber, pepper, lettuce, eggplant, spinach and some of the most widely used tomato cultivars and hybrids in Italy. For unknown reasons, it was not transmitted from tomato to tomato by *Aphis fabae, Macrosiphum euphorbiae* and *Myzus persicae*. In tomato plants, CMV–ST did spread systemically, reaching a maximum concentration within seven days after inoculation, to decrease rapidly to very low levels that were maintained for at least five months after inoculation. The maximum and minimum CMV–ST titre was about four times lower than that of the parental isolate CMV–Tfn.

Apparently, CMV–ST was able to prevent infection by CMV–PG, for neither the challenging strain nor new pseudorecombinants were recovered by mechanical or aphid transmission from challenged plants. However, since CMV–ST is a pseudorecombinant strain, the possibility was tested that parental strains could be reconstituted, or that other reassortants could arise upon infection of the preimmunized plants. The results showed that parental strains could indeed be reconstituted (or outcompete the pseudorecombinant) only when the pseudorecombinant and the challenging strain were inoculated in mixture or onto two different cotyledons at the same time. Such reconstitution was not observed when plants were challenged three days after preimmunization, either on preimmunized cotyledons or on true leaves. In the latter case, the challenging virus seemed totally unable to initiate infection, as suggested by the failure of its detection with routine diagnostic methods, and lack of multiplication following six subinoculations onto tomato seedlings. The possibility that true recombination could take place between satRNAs of the protecting and challenging strains was excluded by RNAse Protection Assay (RPA), both in preimmunized plants and in their progeny (F. Cillo, L. Barbarossa and D. Gallitelli, unpublished results). This possibility is now being investigated for genomic RNAs.

Experimental fields with tomato seedlings preimmunized with CMV–ST were established in 1994 and 1995 in Apulia. Seedlings were transplanted to the field three weeks after preimmunization, when the virus titre was already stabilized at low levels. In both years, natural CMV infections were too low to evaluate the protective effect of CMV–ST. The trial, however, provided the following information: (i) reduction in fruit yield caused by CMV–ST infection was of about

5%; (ii) fruit ripening and colour were more homogeneous than in control plants; (iii) no statistical differences in pH/soluble solids were observed between control and preimmunized plants; (iv) the nucleotide sequence of ten Tfn–satRNA variants amplified by PCR from tissues collected from preimmunized plants just before harvesting (middle of August) showed two base changes in some clones (T–172 was changed to C, and C–201 was changed to G); (v) in agreement with its unexplained lack of aphid transmissibility, CMV–ST was not recovered from weeds either growing within or adjacent to fields (Cillo *et al.*, 1995).

From the experience with cross–protection to the planning for transgenic tomato plants expressing a benign satRNA variant

Considered together, the results reported so far lead to the following conclusions: i) under natural conditions the satRNA variants seem to have a more stable biological expression than in the laboratory; ii) viral genomic control accounts for most of this stability; iii) the accuracy in selecting the most appropriate helper strain might reduce risks that a benign satRNA could convert into a molecule with deleterious functions.

It has been reported that the biological expression of certain satRNA variants maps specifically on genomic RNAs 1 and 2 of the helper strain (Sleat *et al.*, 1994). This was also the case with Tfn–satRNA. In the presence of its natural helper strain (CMV–Tfn) the satRNA co–determines a transitory stunting in tomato seedlings inoculated at the cotyledon stage (Crescenzi *et al.*, 1993a). This biological trait seems to map specifically on RNAs 1 and 2 of CMV–Tfn, since with heterologous CMV strains the Tfn satRNA attenuates symptoms elicited in tomato. This was particularly true when replication of Tfn–satRNA was supported by CMV–S, a mild strain with a powerful capacity to support symptom modulation. For example, the tomato lethal necrosis disease was converted into a stunting phenotype when CMV–S was supplied as the helper for the necrogenic PG–satRNA (D. Gallitelli, unpublished).

Our experience with transgenic plants was initiated approximately in the middle of 1995 with a program for transformation of UC82 tomato plants with the following constructs:

1) Tfn–satRNA expressed under the tissue–specific promoter gPAL2, aimed to compartimentalize the satRNA in the leaf epidermis, the tissue where aphids probe;
2) Tfn–satRNA and Tfn–satRNA plus Tfn–CP expressed under the classical CaMV 35S promoter to serve as control (Yie *et al.*, 1993);
3) Tfn–satRNA plus the functional proteins 1a and 2a of CMV–S, aimed at controlling the biological expression of the satRNA variant and to suppress replication of the challenging strain.

Conclusions

The recent paper published by Palukaitis and Roossinck (1996) renewed alarm concerning the risks associated with a deliberate use of benign CMV satRNA molecules as biocontrol agents, and in their comments, Tepfer and Jacquemond (1996) envisaged the possible end of projects for the commercial use of benign CMV–satRNAs. However, several considerations were made by the latter authors, leaving open a window for the possibility that strong enough mechanisms could operate in nature to counterselect benign satRNA molecules that acquire deleterious functions. Although we give credit to the model proposed by Palukaitis and Roossinck (1996), and in principle agree with all cautionary concepts, we must carefully look at that window, considering in particular that the data obtained from nature do not seem to fit completely greenhouse experiments.

Acknowledgements

This work was supported in part by a grant of MIRAAF, D.M. 121/7240/96 del 14.3.96, prog. 171 del Piano nazionale "Biotecnologie Vegetali", dal titolo " Ottenimento di piante di pomodoro con livelli utili di resistenza non convenzionale al virus del mosaico del cetriolo (CMV)".

References

Cillo F, Tamba ML, Borgatti S, Cerato C, Di Candilo M, Barbarossa L, Di Franco A, Grieco F, Mileti V, Gallitelli D (1995) Biological properties of a satellite–containing pseudorecombinant of cucumber mosaic virus and its use for tomato protection in Italy. In: Abstracts of the 6th International Plant Virus Epidemiology Symposium, Ma'ale Hachamisha, Israel, p 47

Collmer CW, Howell SH (1992) Role of satellite RNA in the expression of symptoms caused by plant viruses. Annu Rev Phytopath 30:419–422

Crescenzi A, Barbarossa L, Cillo F, Di Franco A, Vovlas N, Gallitelli D (1993) Role of cucumber mosaic virus and its satellite RNA in the etiology of tomato fruit necrosis in Italy. Arch Virol 131:321–333

Crescenzi A, Barbarossa L, Gallitelli D, Martelli GP (1993) Cucumber mosaic cucumovirus populations in Italy under natural epidemic conditions and after a satellite–mediated protection test. Plant Dis 77:28–33

Gallitelli D, Martelli GP, Gebre–Selassie K, Marchoux G (1995) Progress in the biological and molecular studies of some important viruses of solanaceae in the Mediterr Acta Hortic 412:503–514

Grieco F, La Nave C, Gallitelli D (1997) Evolutionary dynamics of cucumber mosaic virus satellite RNA during natural epidemics in Italy. Virology 229: 166–174

Kaper JM (1993) Satellite–mediated symptom modulation: an emerging technology for the biological control of viral crop disease. Microb Releases 2:1–9

Kaper JM (1995). Role of satellites in viral pathogenesis: nested parasitic nucleic acids competing for expression. In: Singh RP, Singh U, Kohmoto K (ed) Pathogenesis and host specificity in plant diseases, Histopathological, biochemical, genetic and molecular bases, vol III Viruses and Viroids. Pergamon Press, p 377–392

Palukaitis P, Roossinck MJ (1996) Spontaneous change of a benign satellite RNA of cucumber mosaic virus to a pathogenic variant. Nature Biotechnology 14:1264–1268

Roossinck MJ, Sleat D, Palukaitis P (1992) Satellite RNAs of plant viruses, Structures and biological effects. Microbiol Rev 56:265–279

Rybicki EP (1995) Family *Bromoviridae*. In: Murphy FA, Fauquet CM, Bishop DH, Ghabrial SA, Jarvis AW, Martelli GP, Mayo MA, Summers MD (ed) Virus Taxonomy, Springer-Verlag, Wien, New York, pp 453–457

Sleat DE, Zhang L Palukaitis P (1994) Mapping determinants within cucumber mosaic virus and its satellite RNA for the induction of necrosis in tomato plants. Molec Plant-Microbe Interact 2:189–195

Tepfer M (1993) Viral genes and transgenic plants. what are the potential environmental risks? Bio/Technology 11: 1125–1132

Tepfer M, Jacquemond M (1996) Sleeping satellites: A risky prospect. Nature Biotechnology 14:1226

Tien P, Wu GS (1991) Satellite RNA for the biocontrol of plant disease. Adv Virus Res 39:321–339

Wilson TMA (1993) Strategies to protect crop plants against viruses: Pathogen-derived resistance blossoms. Proc Natl Acad Sci USA 90:3143–3141

Yie Y, Tien P (1993) Plant virus satellite and their role in engineering resistance to virus disease. Semin Virol 4:363–368

Ecological Impact of Transgenic Virus–Resistance in Crop, Weed, and Wild Plant Populations (Due to Potential Alterations of Plant Invasiveness)

Detlef Bartsch

Department of Biology V, Ecology, Ecochemistry and Ecotoxicology, Aachen University of Technology - RWTH, Worringerweg 1, 52056 Aachen, Germany

Abstract

Large–scale cultivation of plants expressing viral genes raises several ecological questions, including: where are sexually compatible cultivar, weed, and wild plant populations of the virus–influenced plant? where is the virus (naturally) present and how long has the virus been there? which populations might the virus invade in the future? how much is the virus (or the potential recombinant and heterologous encapsidated virus relative) able to decrease the fitness of susceptible plants? what is the potential velocity of gene flow *to* and gene introgression *into* weed and wild plant populations? For example, beet (*Beta vulgaris*) is a crop of particular concern regarding invasiveness because it commonly hybridises with annual forms of wild beet. There is still a significant lack of knowledge about the potential ecological impact of transgenic plants expressing viral sequences. Several disadvantages, but also some advantages, of the spreading of virus resistance genes to non–target plant populations are possible. Virus resistance would be most likely to cause "ecological release" in wild populations if the disease is widespread and has significant effects on plant fitness. In the future, plant breeding activities will develop improved transgenic high–performance cultivars. An ecological inventory and monitoring of wild plant habitats would help to assess the risks of transgenic gene introgression.

Introduction: Potential ecological impact

Transgenic plants are considered a potential risk if they contain a trait that confers a significant fitness advantage in natural situations. In order to set endpoints for "ecological impact", our aim should be the "*conservation of all plant and animal species, including those which have developed together with man, in survivable populations, in their natural communities*". This aim of human society is also agreed on by most scientists and should also apply to new evolutionary developments.

One object of the United Nations Environment Programme is the conservation and sustainable use of biological diversity. Effort should be taken to establish or maintain the means to regulate, manage or control the risks associated with the use and release of living genetically modified organisms, which are likely to have adverse environmental impacts that could affect the conservation and sustainable use of biological diversity. Wild relatives of domesticated or cultivated species are important objects of conservation.

Sectors for the assessment of the ecological impact

Transgenic plants encoding viral sequences will primarily behave in nature like their non-transgenic equivalents or their relatives. Ecological interactions must therefore be viewed in comparison to known naturally existing phenomena, e.g. gene flow, even if there is still a lack in basic scientific knowledge.

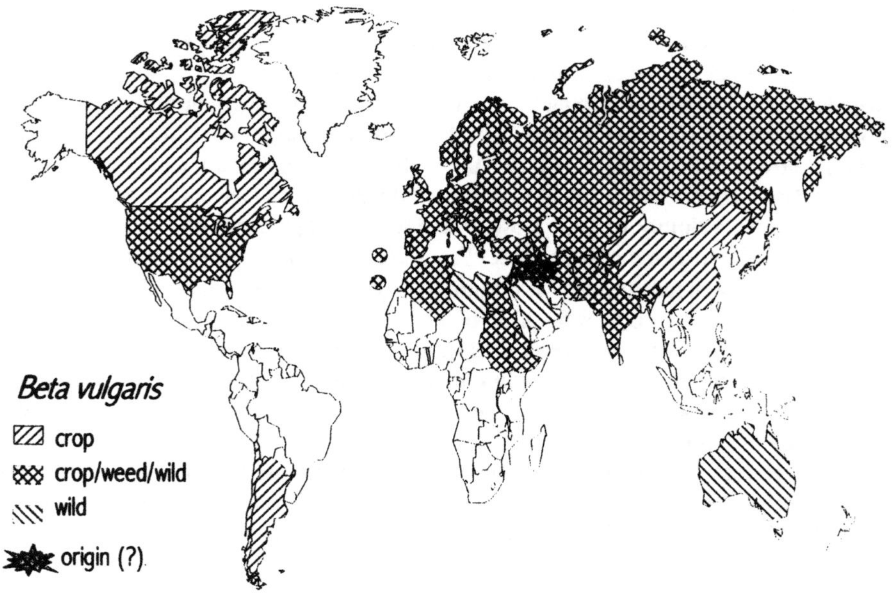

Fig. 1. World-wide distribution (on base of states) of crop, weed and wild varieties of *Beta vulgaris*:. Data are extracted from George (1980), Frese (1991), and the United States Department of Agriculture database (USDA-ARS GRIN, wep page: gopher://gopher.ars-grin.gov:70/77/bin/gateways/statstax?Beta vulgaris).

Agricultural and natural habitats

Where are locations of sexually compatible cultivar, weed, and wild plant populations of the virus-influenced plant species? For instance, the distribution of cultivated beet, *Beta vulgaris*, is ecologically limited, due to its low competitiveness and susceptibility to a number of plant pathogens and herbivores. Therefore, an 'escape' of transgenic attributes of sugar beet is more likely via gene flow into populations of beet relatives than the escape of transgenic sugar beet itself. It is important to emphasise that all cultivated and wild varieties of beet (sugar beet, swiss chard, red beet or fodder beet) belong to the same species. This implies that there is no barrier for outcrossing of fixed genetic qualities within this species (Bartsch and Pohl-Orf, 1996). Meanwhile, beets have reached a broad world-wide distribution (Fig.1). With the introduction of new genes that do not normally occur in this species, either into cultivated plants or into crossable relatives, the evolutionary principle of mutation and selection is accelerated in a way that can not be expected under natural conditions. Thus it is very important to consider, case-by-case, the risks that such a manipulation could entail (Crawley, 1993; Dale, 1993; Parker and Bartsch, 1996).

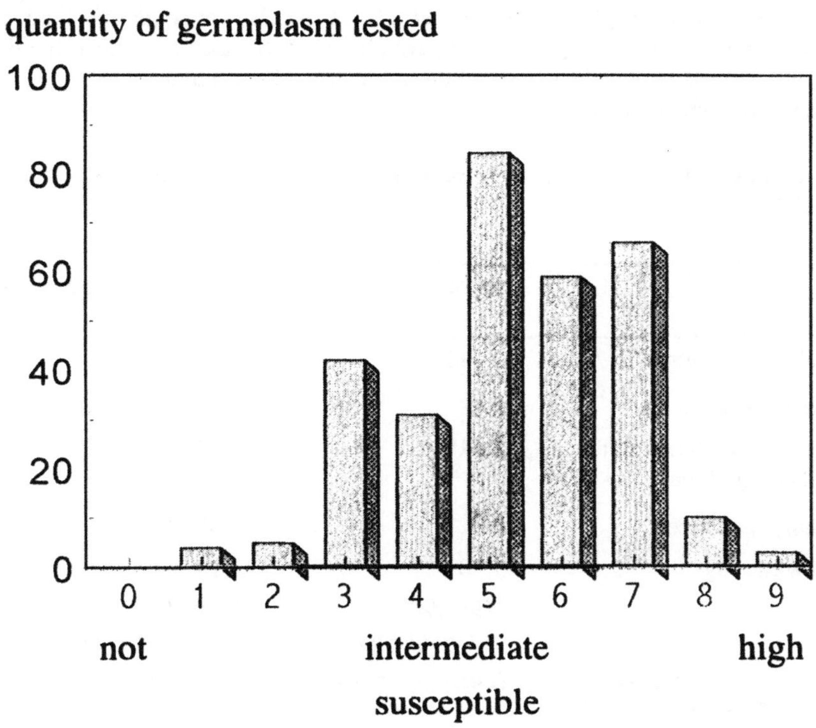

Fig. 2. Susceptibility of *Beta vulgaris* germplasm to BNYVV from various world-wide locations. Data are extracted from the United States Department of Agriculture database (USDA-ARS GRIN, web page: http://www.ars-grin.gov/cgi-bin/npgs/html/ codes.pl?49085).

Virus location and virus-host coevolution

Where is the virus (naturally) present and how long has the virus been there? The trait expressed by the transgenic beet in our studies (Mannerlöf *et al* ., 1996), is resistance to rhizomania caused by beet necrotic yellow vein furovirus (BNYVV), a disease which has spread through the sugar beet fields of Europe, USA, Japan, and

China (Cooper and Asher, 1988; Asher, 1993). The virus is transmitted via the fungus *Polymyxa betae* Keskin (Tamada, 1975; Abe and Tamada, 1986). The disease leads to decreased sugar beet yields, and a loss of up to 30% sugar content (Giunchedi *et al.*, 1987). In some places in Italy where the disease is very severe, sugar beet production has been totally curtailed (Faccioli and Giunchedi, 1974). Whole regions of the USA have been taken out of sugar beet production because of rhizomania.

Virus epidemiology

Which populations might be invaded by the virus or the new genes in the future? Little is known about rhizomania in wild beet habitats and about its ecological impact on wild beet fitness (Bartsch *et al.*, 1996). However, the fact that in conventional plant breeding, wild beet germplasm can be used to achieve (partial) tolerance against against the virus (Whitney 1989), could be a hint for an ecological impact of rhizomania resistance (Fig. 2).

Table 1. Performance of transgenic plants in comparison to the non-transgenic control *Beta vulgaris* at high BNYVV infestation level

	sugar beet hybrids	sugar beet x spinach beet hybrids
Seedling emergence	1.11[1]	n.d.
Biomass production	**1.61***	**1.18***
Biomass production under weed competition	**1.61***	**1.23***
Bolting resistance+	n.d.	**5.01*+**
Overwintering	1.00	1.16
	(1995/96 – strong winter)	*(1996/97 – 'normal' winter)*
Final rate of increase (seeds produced per seeds sown)	*(intended for 1998)*	*(intended for 1999)*

[1] All data of field experiments are expressed as the quotient transgenic versus control. (n.d. = not detected, * statistically significant effect, + effect probably due to somaclonal variation)

Fitness impact

How much is the virus (or the potential recombinant and heterologous encapsidated virus relative) able to decrease the fitness of susceptible plants? Our investigations simulate an outcrossing of the genetically engineered attributes to the sugar beet relative, swiss chard, as a model, and measure the performance of the hybrids under conditions in which BNYVV-resistance might be advantageous (Table 1). No heteroencapsidation or other non–intended side effects are reported in beets expressing the coat protein of BNYVV (König *et al.*, 1995).

Plant population genetics

What is the potential velocity of gene flow *to* and gene introgression *into* weed and wild plant populations? For example, *Beta vulgaris* is a crop of particular concern regarding invasiveness, because it commonly hybridises with annual forms of wild beet. With respect to recent investigations (Boudry *et al.*, 1993; Santoni and Berville, 1992), where markers of cultivated beets have been found in wild beet populations, we conclude that there will be exchange and introgression of the new genetic information in wild beet populations (Bartsch and Schmidt, 1997). If there is a release of transgenic sugar beets on a large scale, we have to expect dispersal of the new attributes.

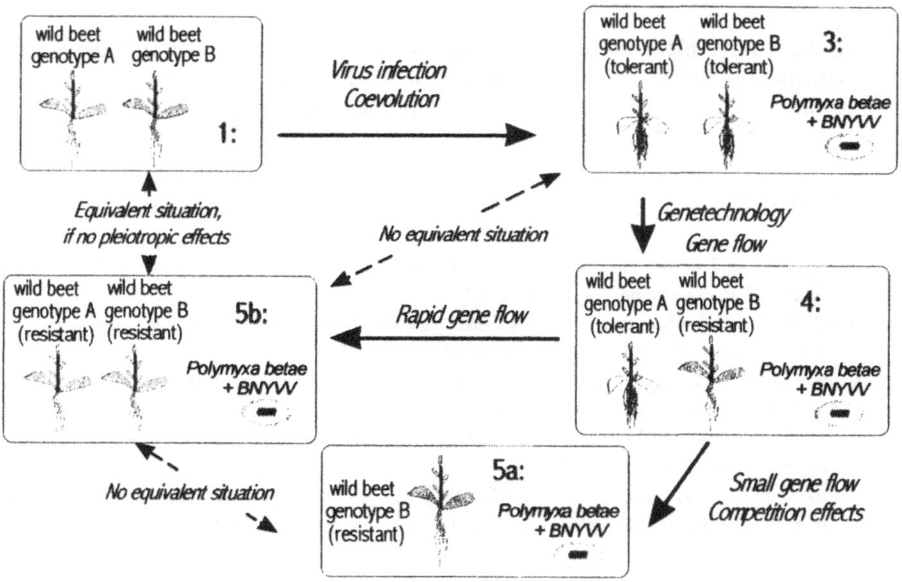

Fig. 3. Theoretical model of potential unwanted effects due to gene flow of BNYV–resistance from cultivar to wild plant populations of *Beta vulgaris:*.The effects depend on the initial step in the wild beet habitat (step 1 or step 3; step 2 is not shown in this figure) and will probably result in one of two different final steps (5a or 5b). Step 5a represents a loss of biodiversity. Also, tolerant plants reproduce the virus. Infected plants are shown with bleached and withered leaves. *Polymyxa betae* is the vector of the soil–born rhizomania disease caused by BNYVV.

Assessment

There is still a significant lack of knowledge about the potential ecological impact of transgenic plants expressing viral sequences. Several disadvantages, but also some advantages, of the spreading of virus resistance genes to non-target plant populations are possible. Virus resistance would be most likely to cause "ecological release" in wild populations if the disease is widespread and has significant effects on plant fitness. Ecological implications due to the introduction and spreading of transgenic hybrids might therefore only be observable in natural beet habitats with high level of BNYVV, where susceptible *Beta vulgaris* genotypes grow (Fig. 3). Under these circumstances, introgression of transgenic resistance genes would surely alter the natural genetic constitution of wild beet populations.

Prospects

In the future, plant breeding activities will develop improved transgenic high-performance cultivars. An ecological inventory and monitoring of wild plant habitats would help to assess the risks of transgenic gene introgression.

References

Abe H, Tamada T (1986) Association of beet necrotic yellow vein virus with isolates of *Polymyxa betae* Keskin. Ann Phytopathol Soc Japan 52:235–247

Asher MJC (1993) Rhizomania. in: DA Cooke and RK Scott (ed) The sugar beet crop: science into practice. Chapmann and Hall, London, pp 311–346

Bartsch D, Pohl-Orf M (1996) Ecological aspects of transgenic sugar beet: Transfer and expression of herbicide resistance in hybrids with wild beets. Euphytica 91: 55–58

Bartsch D, Schmidt M (1997) Influence of sugar beet breeding on populations of *Beta vulgaris* ssp. *maritima* in Italy. J Veget Sci 8:81–84

Bartsch D, Schmidt M, Pohl-Orf M, Haag C, Schuphan I (1996) Competitiveness of transgenic sugar beet resistant to beet necrotic yellow vein virus and potential impact on wild beet populations. Mol Ecol 5:199–205

Boudry P, Mörchen M, Saumitou-Laprade P, Vernet P, Van Dijk H (1993) The origin and evolution of weed beets: consequences for the breeding and release of herbicide-resistant transgenic sugar-beets. Theor Appl Genet 87:471–478

Cooper JI, Asher MJC (1988) *Viruses with fungal vectors*. Association of Applied Biologists, Wellesbourne. 355 pp.

Crawley MJ, Hails RS, Rees M, Kohn D, Buxton J (1993) Ecology of transgenic oilseed rape in natural habitats. Nature 363:620–623

Dale PJ (1993) The release of transgenic plants into agriculture. J Agric Sci 120: 1–5

Faccioli G, Giunchedi L (1974) On the viruses involved on rhizomania diseases of sugar beet in Italy. Phytopathol mediterr 13:28–35

Frese L (1991) Sammlung, Erhaltung und Nutzbarmachung der genetischen Ressourcen von Beta-Rüben (*B. vulgaris* L.) und Wurzelzichorien (*Cichorium intybus* L.) Landbauforschung Völkenrode 41:65–73

George AS (ed) (1980) Flora of Australia, new ed. 4: p277

Giunchedi L, Biaggi M, Poggi Pollini C (1987) Correlation between tolerance and beet necrotic yellow vein virus in sugar beet genotypes. Phytopathol mediterr 26:23–28

König R, Lesemann DE, Maiss E (1995) Attempts to detect heteroencapsidations or other non-intended side effects in transgenic sugarbeet expressing the coat protein gene of beet necrotic yellow vein virus (BNYVV). In: Landsmann J, Casper R (ed) Key Biosafety Aspects of Genetically Modified Organisms - Workshop 10.–11. April 1995 in Braunschweig, Germany, Mitteilungen aus der Biologischen Bundesanstalt für Land- und Forstwirtschaft 309, 71–7

Mannerlöf M, Lennerfors B-L, Tenning P (1996) Reduced titer of BNYVV in transgenic sugar beets expressing the BNYVV coat protein. Euphytica 90:293–299

Parker IM, Bartsch D (1996) Recent advances in ecological biosafety research on the risks of transgenic plants: a trans-continental perspective In: Tomiuk J, Sentker A, Wöhrmann K (ed) Transgenic Organisms - Biological and Social Implications. Birkhäuser-Verlag Basel, 147–161

Santoni S, Berville A (1992) Evidence for gene exchange between sugar beet (*Beta vulgaris* L.) and wild beets: consequences for transgenic sugar beets. Plant Mol Biol 20:578–580

Tamada T (1975) Beet necrotic yellow vein virus. *Commonwealth Mykological Institute and Association of Applied Biologists Description of Plant Viruses,* set 4, No. 144

Whitney ED (1989) Identification, distribution, and testing for resistance to rhizomania in *Beta maritima*. Plant Dis 73:287–290

Risk Assessment of Gene Flow Associated with the Release of Virus–Resistant Transgenic Crop Plants

Marc Fuchs and Dennis Gonsalves
Department of Plant Pathology, Cornell University, New York State Agricultural Experiment Station, Geneva, New York 14456, USA.

Abstract
An important environmental concern associated with the large scale cultivation of virus–resistant transgenic crop plants relates to the possible transfer of virus-derived transgenes into wild relatives through pollen flow. We addressed this issue by using two squash (*Cucurbita pepo* ssp. *ovifera*) species: The transgenic line CZW–3 (*C. pepo* var. *ovifera*) containing the coat protein (CP) genes of cucumber mosaic cucumovirus (CMV), zucchini yellow mosaic potyvirus (ZYMV), and watermelon mosaic virus 2 potyvirus (WMV 2) and its wild relative *C. pepo* var. *texana*. Viral transgene flow was evaluated in the field by studying the successive rate of introgression of the three CP genes from F1 hybrids (= *C. texana* x CZW–3) and their backcrossed progeny (BC 1 = *C. texana* x F1 hybrids) into *C. texana* squash populations. Field plots were maintained under low or high disease pressure by deliberately introducing CMV, ZYMV, and WMV 2 in the latter plots. Introgression occurred under low but not high disease pressure. To assess fitness, test plants (*C. texana*, CZW–3, F1 hybrids and BC 1 progeny) were evaluated for virus resistance, number of fruits produced, and number of viable seeds. Under high disease pressure, all *C. texana* plants became severely infected and produced less than one fruit per plant and no viable seeds. Hybrids of *C. texana* that contained the CP genes, on the other hand, exhibited virus resistance and produced a fair number of fruits with viable seeds. The implications of these data will be discussed.

Introduction

The concept of pathogen–derived resistance (Sanford and Johnston, 1985) has been used extensively and successfully to engineer virus resistance in numerous crops. Coat protein (CP) genes have been used more often than any other viral–derived genes to engineer virus–resistant agronomically important transgenic crop plants (Fuchs *et al.*, 1997a).

Recently, the commercialization of virus–resistant transgenic crop plants has become a reality in the United States. Freedom II, the first virus– and disease-resistant transgenic crop plant released, was deregulated in the U.S. in 1994 (Medley, 1994) and commercialized in the spring of 1995 by the Asgrow Seed Company. Freedom II is a summer squash line that contains the CP genes of zucchini yellow mosaic potyvirus (ZYMV) and watermelon mosaic virus 2 potyvirus (WMV 2). This line is resistant to these two viruses and produces high yields of marketable fruits upon severe infections by ZYMV and WMV 2. Prior to its release, Freedom II was tested experimentally under the code name ZW-20 (Tricoli *et al.*, 1995; Fuchs and Gonsalves, 1995). In 1996, the squash line CZW-3 was the second virus resistant transgenic product to receive exemption status in the U. S. (Acord, 1996). This squash line contains the CP genes of cucumber mosaic cucumovirus (CMV), ZYMV and WMV 2, and is highly resistant to these three aphid-borne viruses (Tricoli et *al.*, 1995). Line CZW-3 produces high yields of marketable fruits upon mixed infections by CMV, ZYMV and WMV 2 (Fuchs *et al.*, unpublished data). In 1996, papaya became the first transgenic fruit crop to be deregulated in the U.S. (Strating, 1996). Deregulated papaya line 55-1 expresses the CP gene of papaya ringspot potyvirus (PRSV) (Fitch *et al.*, 1992). Papaya trees of line 55-1 were completely resistant to PRSV, even after two years in the field in Hawaii (Lius *et al.*, 1997). Squash line CZW-3 and papaya line 55-1 are expected to be released commercially this year.

Other agronomically important virus-resistant transgenic crops are likely to be cleared for commercial use in the near future. Among these is a transgenic cantaloupe line (CZW-30) that contains the CP genes of CMV, ZYMV, and WMV 2, and a transgenic tomato line (TT-5) that expresses the CP gene of CMV. Field experiments have shown that cantaloupe line CZW-30 exhibited resistance to mixed infections by CMV, ZYMV, and/or WMV 2 (Fuchs *et al.*, 1997b). Hybrid plants of line CZW-30 had a 7-fold increase in yield of marketable fruits compared to untransformed plants under conditions of severe disease pressure (Fuchs *et al.*, 1997b). Tomato line TT-5, expressing the CP gene of CMV strain WL, displayed high levels of resistance to several strains from the two CMV subgroups (Xue *et al.*, 1994; Provvidenti and Gonsalves, 1995). Field studies showed that line TT-5 had a 17-fold higher fruit yield than inoculated control plants (Fuchs *et al.*, 1996).

Virus-resistant transgenic crops offer numerous benefits to agriculture. These benefits are of agronomical (increased potential for resistance when host resistance genes have not been identified and for resistance to multiple viruses), economical (increased yield and improved crop quality), epidemiological (reduced sources of viruses for secondary spread and potential to limit epidemics) and environmental (reduced use of insecticides) significance.

Despite the potential benefits that virus-resistant transgenic crop plants offer to agriculture, concerns have been raised with their widespread release into the environment (de Zoeten, 1991; Palukaitis, 1991; Tolin, 1991; Wilson, 1993; Rissler and Mellon, 1993; Tepfer, 1993). One of the major environmental issues relates to the possible escape of viral-derived transgenes from transgenic crops into wild relatives (Hancock *et al.*, 1996). Numerous articles and reviews have been published on the potential environmental risks associated with the commercial release of virus-resistant transgenic crop plants. However, only limited scientific information on safety assessment is available from experiments conducted in the field. Scientific data from field experiments are critical to evaluate the environmental effects associated with the release of virus resistant transgenic crops.

We assessed under field conditions the flow of transgenes from a virus-resistant transgenic crop plant into a free-living relative, by using the following two squash species: The transgenic summer squash (*Cucurbita pepo* ssp. *ovifera* var. *ovifera*) line CZW-3 that contains the CP genes of CMV, ZYMV and WMV 2 and its wild relative (*Cucurbita pepo* spp. *ovifera* var. *texana*) known as Texas gourd and referred to hereafter as *C. texana*. This free-living squash species is found in several states in the south-central part of the U.S. (Wilson *et al.*, 1994). *C. texana* can be a localized weed problem in soybean and cotton fields (Weideman and Templeton, 1988) and can hybridize readily with cultivated squash species. Gene flow has been demonstrated to occur naturally between cultivated and *C. texana* squash and vice versa (Kirkpatrick and Wilson, 1988). Thus, it is conceivable that CP genes will escape from virus-resistant transgenic squash into *C. texana* plants that grow in close proximity to commercial fields of transgenic squash. Results from two year field experiments designed to assess the rate of CP gene introgression and the fitness of wild x transgenic hybrid squash plants will be presented and discussed.

Environmental safety assessment of viral-derived transgene flow

Introgression of CP genes from wild x transgenic hybrid squash into wild squash
The basic objective of these experiments was to monitor CP transgene flow to *C. texana* squash populations over successive growing seasons. To reproduce natural conditions as much as possible, the same field plots were used over time. In 1995, we analyzed the rate of integration of the CP genes from wild x transgenic crop F1 hybrids into a population of wild plants. F1 hybrids (= *C. texana* x CZW-3) were developed by hand crossing transgenic CZW-3 and *C. texana* squash in the greenhouse. For these crosses, CZW-3 was used as staminate parent and *C. texana* as pistillate parent to mimic pollen flow from transgenic into wild squash. In 1996, experiments were carried out in the same fields as in 1995 with the progeny of the plants tested in 1995. During both years, field trials were conducted under low and high disease pressure. Low disease pressure was achieved by transplanting to the field sites uninoculated plants and by quickly roguing plants that became aphid-infected early in the growing season. High disease pressure was achieved by transplanting mechanically inoculated *C. texana* plants to the field sites along with the test plants. These mechanically inoculated plants served as virus source for natural transmissions by indigenous aphid populations.

To assess whether introgression occurred, *C. texana* fruits were collected and their seeds were harvested. The presence of the CP genes was indirectly determined in germinating seeds by detecting the linked neomycin phosphotransferase II (NPT II) protein by enzyme-linked immunosorbent assay (ELISA). Previous experiments had shown that virus resistance correlated well with the expression of the NPT II gene in squash line CZW-3 and any of its progeny, as the CP and the NPT II transgenes are usually co-transferred.

Our results showed that a high number of seeds from *C. texana* fruits collected in 1995 in fields with low disease pressure reacted positively in ELISA for the NPT

II protein, indicating the presence of the transgenes (Table 1). Introgression did not occur under conditions of high disease pressure, because infections by CMV, ZYMV and WMV 2 early in the growing season affected severely the development and growth of *C. texana* plants by impairing their flowering and fruiting potentials. Successive introgression was detected in the 1996 trials in which seedlings from fruits collected in 1995 were used. Again, introgression only occurred in plots with low disease pressure.

Fitness evaluation of wild x transgenic hybrid squash

In parallel to the introgression studies, we assessed in the field the fitness of wild squash that acquired the CP genes either by controlled hand pollination or by introgression. In 1995, we evaluated the resistance of F1 hybrids (= *C. texana* x CZW–3), *C. texana* squash and transgenic CZW–3 squash to infections by CMV, ZYMV and WMV 2. In 1996, we compared the virus resistance of F1 hybrids, the

Table 1: Assessment of CP gene introgression from transgenic *C. texana* x CZW–3 F1 hybrid squash into *C. texana* squash in 1995.

Farms	Fields	Ratio F1 hybrid/ *C. texana* (%)	Disease incidence	n	*C. texana* seed viability(%)[a]	NPT II ELISA No positive/Total[b]	(%)
I, II	A, B	12	low	56	83	466/1260	37
	C, D	4	low	48	87	189/360	30
III, IV	E, F	12	high	56	0	nt	nt
	G, H	4	high	48	0	nt	nt

[a]Seed viability was mesured as the percentage of seeds that germinated by analyzing 10 seeds per fruit and 10 fruits per plant

[b]Number of *C. texana* seeds that reacted positively over the total number of seeds tested. An average of 10 seeds were analyzed per fruit, and 10 fruits per *C. texana* plant

n = number of *C. texana* plants tested

nt = not tested

first backcrossed progeny (BC 1 = *C. texana* x F1 hybrids), *C. texana* squash and transgenic CZW–3 squash. During both years, field trials were conducted under low and high disease pressure. High disease pressure was established by deliberately inoculating a percentage of nontransgenic squash with the three viruses, and low disease pressure was attempted by isolating the field plots from external virus sources. Virus resistance was evaluated by visual monitoring of symptom development and ELISA detection of viruses. Furthermore, fitness was evaluated by counting the number of fruits produced by each genotype and by analyzing their seed viability.

Results from the 1995 field experiments showed that transgenic F1 hybrid plants were symptomless or had local chlorotic lesions, similar to those displayed by the transgenic CZW–3 plants, regardless of the disease incidence (Table 2). In contrast, all nontransgenic hybrid and *C. texana* plants became symptomatic with severe symptoms 3–4 and 9–10 weeks post-planting, under low and high disease pressure, respectively (Table 2). Resistance was confirmed by ELISA. Transgenic F1 hybrids and CZW-3 plants did not react positively in ELISA for CMV, ZYMV and WMV 2, whereas all nontransgenic hybrid and *C. texana* plants gave positive reactions for these three viruses (Table 2). Under low disease pressure in 1995, *C. texana* plants produced significantly more fruits, but smaller in size, than transgenic F1 hybrid and CZW-3 plants. However, transgenic F1 hybrid plants produced more fuits under high disease pressure than nontransgenic hybrid and *C. texana* plants.

Table 2: Fitness evluation of *C. texana*, F1 hybrid, and transgenic CZW-3 squash in 1995.

Genotype[a]	Disease incidence	n	Plants with severe symptoms (%)	ELISA positive plants (%)[b]	Number of fruits/plant	Number of seeds/fruit
Nontransgenic						
C. texana	low	96	100	100	18 ± 6	175 ± 2
	high	41	100	100	<1	<5
C. texana x YS-20	low	96	100	100	3 ± 1	314 ± 17
	high	47	100	100	1 ± 1	123 ± 48
Transgenic						
C. texana x CZW-3	low	96	0	0	5 ± 1	301 ± 30
	high	12	0	0	5 ± 1	292 ± 45
CZW-3	low	96	0	0	2 ± 1	350 ± 42
	high	84	2	2	2 ± 1	375 ± 51

[a]F1 hybrids were obtained by hand pollination in the greenhouse by using transgenic CZW-3 or nontransformed YS-20 squash as staminate parent and *C. texana* squash as pistillate parent

[b]ELISA was done for CMV, ZYMV and WMV 2. The few transgenic CZW-3 plants that showed severe symptoms and reacted positively in ELISA were found to be infected by papaya ringspot potyvirus.

n = number of plants tested

In addition, fruits from transgenic F1 hybrids contained substantially more viable seeds than fruits from nontransgenic hybrid and from *C. texana* plants.

In 1996, similar results were obtained. Hybrid plants (F1 and BC 1) that carried the CP genes displayed virus resistance regardless of the disease pressure and produced significantly more fruits with more viable seeds than *C. texana* plants under high disease pressure. Interestingly, the BC 1 progeny had a similar number of fruits compared to *C. texana* plants under low disease pressure. Also, a few BC

1 progeny (8%) exhibited *C. texana*-type morphology and produced *C. texana*-type fruits.

Discussion

Introgression of CP genes from transgenic squash into compatible free-living *C. texana* squash is likely to occur through pollen flow when the two species are grown in close proximity. Our data indicated that CP transgenes can easily move from *C. texana* x transgenic F1 hybrids and their progeny into a *C. texana* population, thus providing virus resistance to the wild squash. However, we found that gene introgression can be strongly inhibited under conditions of high disease pressure. Hybrid plants (F1 and BC 1) that acquired the CP transgenes exhibited increased fitness over *C. texana* under high disease pressure, displaying virus resistance, producing a higher number of fruits, and more viable seeds. As expected, transgenic hybrid plants (F1 and BC 1) did not appear to have fitness advantages over nontransgenic plants under low disease pressure.

Our fitness studies demonstrated that *C. texana* squash that acquire CP transgenes from F1 or BC 1 hybrid squash through pollen flow will exhibit virus resistance. Two generations of hybridization and backcrossing resulted in a small precentage of virus resistant progeny with phenotypic characteristics of *C. texana*. The questions that remain to be answered are: Will a virus-resistant *C. texana* plant become a significant threat as an invasive weed? Or, are viruses limiting the ability of *C. texana* to become a significant weed? Extensive surveys will need to be done to assess the incidence of viruses in natural populations of *C. texana*. Interestingly, preliminary data suggest that viruses are not prevalent in *C. texana* plants growing in nature.

References

Acord BR (1996) Availability of determination of nonregulated status for squash line genetically engineered for virus resistance. Fed Regist 61:33485–33486

de Zoeten GA (1991) Risk assessment: do we let history repeat itself? Phytopathology 81:585–586

Fitch MMM, Manshardt RM, Gonsalves D, Slightom JL, Sanford JC (1992) Virus resistant papaya plants derived from tissues bombarded with the coat protein gene of papaya ringspot virus. Bio/Technology 10:1466–1472

Fuchs M, Gonsalves D (1995) Resistance of transgenic hybrid squash ZW-20 expressing the coat protein genes of zucchini yellow mosaic virus and watermelon mosaic virus 2 to mixed infections by both potyviruses. Bio/Technology 13:1466–1473

Fuchs M, Provvidenti R, Slightom JL, Gonsalves D (1996) Evaluation of transgenic tomato plants expressing the coat protein gene of cucumber mosaic virus strain WL under field condition. Plant Dis 80:270–275

Fuchs M, Ferreira S, Gonsalves D (1997a) Management of virus diseases by classical and engineered protection. Mol Pl Path On-line [http://www.bspp.org.uk/mppol] 1997/0116fuchs

Fuchs M, McFerson J, Tricoli DM, McMaster JR, Deng RZ, Boeshore ML, Reynolds JF, Russell PF, Quemada HD, Gonsalves D (1997b) Cantaloupe line CZW-30 containing the coat protein genes of cucumber mosaic virus, zucchini yellow mosaic virus, and watermelon mosaic virus 2 is resistant to these three viruses in the field. Mol Breeding (in press)

Hancock JF, Grumet R, Hokanson SC (1996) The opportunity for escape of engineered genes from transgenic crops. HortSci 31:1080–1085

Kirkpatrick KJ, Wilson H (1988) Interspecific gene flow in Cucurbita: C. texana vs C. pepo. Am J Bot 75:519–527

Lius S, Manshardt RM, Fitch MMM, Slightom JL, Sanford JC, Gonsalves D (1997) Pathogen-derived resistance provides papaya with effective protection against papaya ringspot virus. Mol Breeding (in press)

Medley TL (1994) Availability of determination of nonregulated status for virus resistant squash. Fed Regist 59:64187–64188

Palukaitis P (1991) Virus-mediated genetic transfer in plants. In: Levin M, Strauss H (ed) Risk Assessment in Genetic Engineering: Environmental Release of Organisms. McGraw-Hill Inc, New York. p 140–162

Provvidenti R, Gonsalves D (1995) Inheritance of resistance to cucumber mosaic virus in a transgenic tomato line with the coat protein gene of the white leaf strain. J Hered 86:85–88

Rissler J, Mellon M (ed) (1996) The ecological risks of engineered crops. MIT Press, Cambridge, Massachussetts, USA, p 159

Sanford JC, Johnston SA (1985) The concept of parasite-derived resistance-Deriving resistance genes from the parasite's own genome. J Theor Biol 113:395–405

Strating A (1996) Availability of determination of nonregulated status for papaya lines genetically engineered for virus resistance. Fed Regist 61:48663–48664

Tepfer M (1993) Viral genes and transgenic plants. What are the potential environmental risks? Bio/Technology 11:1125–1129

Tolin S (1991) Persistence, establishment, and mitigation of phytopathogenic viruses. In: Levin M, Strauss H (ed) Risk Assessment in Genetic Engineering: Environmental Release of Organisms. McGraw-Hill Inc, New York. p 114–139

Tricoli DM, Carney KJ, Russell PF, McMAster JR, Groff, DW, Hadden KC, Himmell PT, Hubbard JP, Boeshore ML, Reynolds JF, Quemada HD (1995) Field evaluation of transgenic squash containing single or multiple virus coat protein gene constructs for resistance to cucumber mosaic virus, watermelon mosaic virus 2, and zucchini yellow mosaic virus. Bio/Technology 13:1458–1465

Weideman GJ, Templeton GE (1988) Efficacy and soil persistence of *Fusarium solani* f sp *cucurbitae* for control of Texas gourd *(Cucurbitana texana)*. Plant Dis 72:36–38

Wilson TMA (1993) Strategies to protect crop plants against viruses : pathogen-derived resistance blossoms. Proc Natl Acad Sci USA 90:3134–3141

Wilson HD, Lira R, Rodriguez I (1994) Crop-weed gene flow: Cucurbita argyrosperma Huber and C. fraterna LH Bailey (Cucurbitacea). Econ Bot 48:293–300

Xue B, Gonsalves C, Provvidenti R, Slightom JL, Fuchs M, Gonsalves D (1994) Development of transgenic tomato expressing a high level of resistance to cucumber mosaic virus strains of subgroup I and II. Plant Dis 78:1038–1041

Concluding Remarks and Recommendations

Mark Tepfer[1] and Ervin Balázs[2]

[1] Laboratoire de Biologie Cellulaire, INRA–Versailles, 78026 Versailles Cedex, France
[2] Institute for Plant Sciences, Agricultural Biotechnology Center, POB 411, 2101 Gödöllö, Hungary

Traditional plant breeding and the ongoing process of improvement of agricultural practices have allowed enormous progress in recent decades in the control of plant pathogens. Nonetheless, although this may not be apparent to the general public in North America and Europe, viral diseases cause serious crop losses, not only in tropical areas, but also in temperate regions. For instance, in Asia and Africa viral diseases have considerable impact on rice productivity, and the spread of virulent strains of plum pox potyvirus in Europe will seriously compromise production of plums and apricots if resistance genes are not developed. It is thus quite appropriate that any new means for creating virus resistance genes should be thoroughly explored for potential deployment. Considering the novelty of transgenic virus–resistant plants, it is also reasonable that the principle of precaution be applied, and that potential risks be evaluated, so that the decision to deploy on a large scale can be firmly based on an overall risk/benefit analysis.

Most of the potential risk factors associated with transgenic plants rendered virus–resistant by expression of viral sequences are in fact similar or identical to events that occur when non–transgenic plants are cultivated and are infected by viruses. There was a strong consensus at the workshop that for this reason the correct baseline for risk assessment is the current situation in the absence of transgenic plants. For instance, virus recombination is a well known feature of virus evolution (see chapters by Gibbs *et al.*, Candresse *et al.*). However, only in recent years has sequencing information made it clear that recombination has worked on viruses at all taxonomic levels, and this has led to increased interest in the detection of recombination events and in the analysis of the effects of recombination (see chapters by Bujarski; Simon *et al.*). It has been shown that recombination can occur between viral sequences expressed from a transgene and a infecting virus (see chapters by Allison *et al.*, Wintermantel *et al.*; Jakab *et al.*). Thus, the appropriate risk assessment question is whether recombination in transgenic plants presents risks in addition to those due to recombination in non–transgenic ones. Similarly, the baseline for potential risks associated with heterologous encapsidation is represented by what already occurs in plants infected with more than one virus, which is a frequent occurrence, and potential risks associated with transmission of virus–derived resistance genes to wild plant species should be compared to that due to transmission of natural virus resistance genes.

It has on occasion been maintained that laboratory studies are of little or no relevance to the only situation where risk is pertinent, the open field. The results presented at this workshop abundantly demonstrated that laboratory studies are very valuable indeed. For instance, laboratory studies on the frequency of recombination in transgenic plants (see chapters by Allison *et al.*; Wintermantel *et al.*) make it possible to design genes whose transcripts will be less apt to recombine with an infecting virus. Similarly, laboratory studies have made it possible to modify coat protein genes to reduce potential risks associated with heterologous encapsidation (see chapter by Maiss *et al.*). More generally, laboratory studies concerning the mechanisms underlying either resistance or potential factors of risk will make it possible to carry out more pertinent studies in the field.

The workshop in Gödöllö was also a unique opportunity for researchers to develop a consensus on priorities for future research.
- In studies of recombination in transgenic plants, now that it is clear that this does occur in plants expressing sequences from several types of viruses (see chapters by Allison *et al.;* Wintermantel *et al.;* Jakab *et al.*), it will be appropriate to focus on two types of questions:
- One is a question of baseline information: is the frequency of recombination in transgenic plants greater than that in plants infected with two or more viruses?
- The other is more a question of consequences: since recombination does occur, what is the outcome? are there changes in host response (symptomatology), or changes in host range? Concerning symptomatology and host range, we already have some elements of response (see chapters by Allison *et al.*, Wintermantel *et al.*, Jacquemond *et al.*). An area in which we have much less information is: if recombinant strains do appear, how well will they compete with the parental strains? It is reasonable to predict that most recombinants will be at a disadvantage, but how general is this? In particular, if recombination leads to changes in virus host range, this would be an instance where recombinants would be expected to have a clear competitive advantage. It was pointed out that our understanding of the determinants of host range are still quite rudimentary.
- Among virus–virus interactions, heterologous encapsidation is beginning to be well understood, but understanding is more rudimentary concerning the potential for synergistic effects (see chapter by Palukaitis and Kaplan), or concerning the effects of infection of transgenic plants with more than one virus (see chapter by Jakab *et al.*). Here further studies are also clearly merited.
- Potential risks associated with CMV satRNA genes make their future commercial use less likely, even though no adverse effects were reported when they were used on a large scale in the People's Republic of China. However, interesting questions remain concerning factors controlling the epidemiology of satRNAs occurring in nature, which could be addressed by a molecular epidemiology approach (see chapter by Gallitelli *et al.*).
- There is also a clear need for research on effects at the ecosystem level in at least two domains.
- Since it has been shown that viral resistance genes can be transmitted from certain crops to wild relatives (see chapters by Bartsch; Fuchs and Gonsalves), can this confer an advantage to the wild species, and even enhance their weediness? Although as mentioned above this question can be raised with plants with conventional resistance genes, transgenic plants could be a particularly useful tool in such studies. This is an area where currently very little is known.
- Molecular techniques for fine-scale studies of the composition of virus populations have come of age (see chapter by Garcia-Arenal *et al.*). It would thus be timely to apply these techniques to situations in which transgenic plants are cultivated, in order to evaluate possible effects on the composition of virus populations. For instance, it would be of great interest to compare, in the field, the

prevalence of recombinant genomes within viral populations in the presence and absence of transgenic plants. Countries where large-scale commercial releases of plants expressing viral sequences are current, such as the People's Republic of China and the US, provide particularly interesting areas for study.

There was also considerable discussion of monitoring associated with releases of plants expressing viral genes. Although the molecular epidemiology studies mentioned above are a form of monitoring, this clearly cannot be carried out in association with the majority of field tests. If some form of observational monitoring is necessary, how should this be carried out and by whom? If agricultural workers themselves will have to be the front line of any large-scale monitoring, where does it go from there? In many countries there is a worrisome and regrettable tendency to train fewer plant pathologists and particularly those working in the field. Classical field pathologists will be the logical relay in any monitoring scheme.

Although this begins to depart from strictly scientific subjects, there was also an interesting discussion, mostly based on anecdotal information, of the state of public acceptance of transgenic plants, and in particular those expressing viral genes. It was interesting to note that scientists working in the field are in quite good agreement, as shown by the group that met in Gödöllö. Nonetheless, there are clear differences between the situation in North America, where public acceptance is generally good, and that in Europe, where the public seems to become increasingly concerned, in particular due to unrelated issues of food safety, e.g. those concerning spongiform encephalopathies. On both continents, the public is generally under-informed of potential benefits that can be expected from virus-resistant plants, and poorly informed as to the scientific evaluation of potential risks. Better education in schools and more complete teacher training would contribute to long-term improvements of this situation. What is the role of the scientist? Can we, or should we contribute more to informing the public, and if so, how?

Index of Viruses

Index of Plants

Printing: Druck Partner Rübelmann, Hemsbach
Binding: Buchbinderei Schaumann, Darmstadt

GPSR Compliance

The European Union's (EU) General Product Safety Regulation (GPSR) is a set of rules that requires consumer products to be safe and our obligations to ensure this.

If you have any concerns about our products, you can contact us on ProductSafety@springernature.com

In case Publisher is established outside the EU, the EU authorized representative is:

Springer Nature Customer Service Center GmbH
Europaplatz 3
69115 Heidelberg, Germany

Batch number: 10682215

Printed by Printforce, the Netherlands

காற்றின் நுண்ணுறவு

ஆலோன் மகரி

Made with ♥ on the Notion Press Platform
www.notionpress.com